MW01380526

WRITERS REPUBLIC

THE TRANSITION

CONCEPTS AND RESOURCES TO HELP YOU HAVE A SUCCESSFUL TRANSITION TO CIVILIAN LIFE.

BRIAN DALE

WRITERS REPUBLIC L.L.C.
515 Summit Ave. Unit R1
Union City, NJ 07087, USA

Website: *www.writersrepublic.com*
Hotline: *1-877-656-6838*
Email: *info@writersrepublic.com*

Ordering Information:
Quantity sales. Special discounts are available on quantity purchases by corporations, associations, and others. For details, contact the publisher at the address above.

Library of Congress Control Number: 2023917539
ISBN-13: 979-8-89100-318-7 [Paperback Edition]
979-8-89100-319-4 [Hardback Edition]
979-8-89100-320-0 [Digital Edition]

Rev. date: 10/12/2023

I would like to dedicate this book to my beautiful wife.
Thank you for your support.

CONTENTS

LIST OF ABBREVIATIONS

BDD: Benefits Delivery at Discharge is a program that allows service members to apply for Veterans Affairs benefits before separating from the military.

BVA: The Board of Veterans' Appeals is an administrative body that reviews appeals from veterans who disagree with decisions made by the Department of Veterans Affairs regarding their benefits claims.

CBT: Cognitive Behavioral Therapy is a psychotherapeutic approach that focuses on addressing the connection between thoughts, feelings, and behaviors to promote positive change and improve mental well-being.

CRDP: Concurrent Retirement and Disability Pay is a program that allows military retirees with service-connected disabilities to receive both their retirement pension and disability compensation concurrently, providing them with increased financial support.

DAV: Disabled American Veterans is a nonprofit organization dedicated to providing support, advocacy, and resources for disabled veterans and their families, assisting them in accessing the benefits and services they deserve.

DIC: Dependency and Indemnity Compensation is a benefit program provided by the Department of Veterans Affairs that provides monthly financial support to eligible survivors of military service members and veterans who died as a result of their service-connected disabilities.

DRO: Decision Review Officer is an experienced official within the Department of Veterans Affairs who conducts a review of appeals and reconsiderations of benefit claims to ensure fair and accurate decisions are made.

DVOP: Disabled Veterans Outreach Program is a program administered by the Department of Labor that offers employment assistance, job training, and career guidance specifically tailored to meet the needs of disabled veterans.

ERG: Employee Resource Groups are voluntary, employee-led groups within an organization that provide support, networking, and resources for specific communities or affinity groups, fostering inclusivity and promoting diversity within the workplace.

GPD: Grant and Per Diem is a program administered by the Department of Veterans Affairs that provides funding to organizations that offer transitional housing and supportive services to homeless veterans, helping them secure stable housing and regain self-sufficiency.

HUD: Housing and Urban Development is a US government agency responsible for developing and implementing policies and programs related to housing, community development, and urban affairs to improve the quality of housing and promote sustainable communities.

HUD-VASH: Housing and Urban Development US Department of Veterans Affairs (VA) Supportive Housing is a collaborative program that combines rental assistance vouchers from HUD with case management and clinical services from the VA to provide housing stability and support for homeless veterans.

IU: Individual Unemployability is a disability benefits program provided by the Department of Veterans Affairs that provides financial compensation at the 100 percent disability rate to veterans who are unable to secure or maintain substantial gainful employment due to service-connected disabilities.

LVER: Local Veterans Employment Representatives are professionals within state employment agencies who specialize in assisting veterans with employment services, job placement, and connecting them with job opportunities in their local communities.

NOD: Notice of Disagreement is a formal written communication submitted by a veteran to the Department of Veterans Affairs to express disagreement with a decision made regarding their claim for benefits, initiating the appeals process.

OVBD: Office of Veterans Business Development is a branch of the US Small Business Administration that provides resources, support, and advocacy for veteran entrepreneurs, assisting them in starting, growing, and succeeding in their small businesses.

PHA: Public Housing Agencies are government entities at the local or regional level that administer public housing programs and provide affordable housing options for low-income individuals and families in need of housing assistance.

PTSD: Post-Traumatic Stress Disorder is a mental health condition that can develop after experiencing or witnessing a traumatic event, characterized by symptoms such as intrusive memories, nightmares, hypervigilance, and avoidance of triggers.

SAH: Specially Adapted Housing is a program offered by the Department of Veterans Affairs that provides eligible disabled veterans with financial assistance to modify or construct homes to accommodate their specific disabilities and enhance their independence.

SBA: Small Business Administration is a US government agency that supports and assists small businesses through loan programs, business counseling, training resources, and advocacy, fostering entrepreneurship and economic growth.

SHA: Special Home Adaption is a program provided by the Department of Veterans Affairs that offers financial assistance to eligible veterans with certain service-connected disabilities to adapt their homes for greater accessibility and independence.

SMC: Special Monthly Compensation is a program administered by the Department of Veterans Affairs that provides additional financial compensation to veterans and service members who have severe disabilities or require assistance with activities of daily living due to their service-connected conditions.

SNAP: Supplemental Nutrition Assistance Program is a federal assistance program that provides eligible low-income individuals and families with electronic benefit transfer (EBT) cards to purchase food and improve their nutritional well-being.

SSVF: Supportive Services for Veteran Families is a program offered by the Department of Veterans Affairs that provides financial assistance and supportive services to very low-income veteran families at risk of homelessness or experiencing homelessness, aiming to promote housing stability and self-sufficiency.

TA: Tuition Assistance is a program provided by the Department of Defense that offers financial support to active duty service members, allowing them to pursue education and training to further their personal and professional development while serving in the military.

TAP: Transition Assistance Program is a comprehensive program that helps service members and their families successfully transition from military to civilian life by providing information, resources, and support in areas such as employment, education, and health care.

TBI: Traumatic Brain Injury is a condition caused by a sudden blow or jolts to the head, resulting in physical, cognitive, and psychological impairments; and it can range from mild to severe in its effects.

VA: US Department of Veterans Affairs is a federal agency that provides comprehensive health-care services, benefits, and support to veterans and their dependents, aiming to fulfill the nation's commitment to those who have served in the armed forces.

VASH: US Department of Veterans Affairs Supportive Housing is a collaborative effort between the Department of Housing and Urban Development (HUD) and the VA that combines rental assistance vouchers with comprehensive case management services to help homeless veterans secure and maintain stable housing.

VASRD: VA's Schedule for Rating Disabilities is a comprehensive guide that outlines the criteria and rating schedule used to evaluate and assign disability ratings to veterans based on the severity of their service-connected conditions.

VBA: Veterans Board of Appeals is an administrative body within the US Department of Veterans Affairs responsible for conducting hearings and making decisions on appeals filed by veterans seeking to challenge previous decisions related to their claims for benefits.

VBOC: Veterans Business Outreach Centers are resource centers funded by the US Small Business Administration that provide support, training, and counseling to veteran entrepreneurs looking to start or expand their small businesses.

VEP: Veterans Entrepreneurship Program is an educational initiative that offers disabled veterans the opportunity to develop their entrepreneurial skills through training, mentorship, and networking opportunities to facilitate the successful launch and growth of their businesses.

VFW: Veterans of Foreign Wars is a nonprofit veterans service organization that provides advocacy, support, and camaraderie to veterans, service members, and their families, while also promoting veterans' rights and well-being.

VHA: Veterans Health Administration is the largest integrated health-care system in the United States, providing comprehensive medical services and support to eligible veterans, with a focus on delivering high-quality and patient-centered care.

VR&C: Vocational Rehabilitation and Counseling is a program offered by the Department of Veterans Affairs that provides eligible veterans with career counseling, training, education, and employment support to enhance their skills and facilitate their successful transition into the civilian workforce.

VR&E: Vocational Rehabilitation and Employment is a program administered by the Department of Veterans Affairs that assists disabled veterans in overcoming barriers to employment by providing vocational rehabilitation services, training, and job placement assistance.

VSO: Veterans Service Organizations are nonprofit organizations that advocate for and provide various support services, benefits assistance, and community engagement opportunities to veterans and their families.

PROLOGUE

The transition from military service to civilian life is a monumental and transformative journey—a journey that marks the end of one chapter and the beginning of another. It is a transition that is unique to each individual, yet shared by a community bound together by a common oath and a shared sense of duty.

For those who have dedicated their lives to serving their countries, the challenges and uncertainties that come with transitioning can be overwhelming. The familiar rhythms of military life are replaced with new routines, new expectations, and a new set of challenges to overcome. It is a time of profound change, where identities are reshaped and new paths must be forged.

In this book, we embark on a voyage of discovery, support, and empowerment. Together, we will explore the intricacies of the transition process, delving into the practical aspects as well as the emotional and psychological dimensions. We will navigate the uncharted waters of career exploration, education, financial management, and health and wellness.

This book is not just a guide; it is a beacon of hope, a companion, and a source of strength. It is designed to provide you, the transitioning veteran, with the tools, resources, and knowledge to navigate the challenges that lie ahead. It is a testament to the resilience, adaptability, and inherent skills that you have acquired during your service.

As we embark on this collective journey, let us remember that we are not alone. Our brothers and sisters, our fellow veterans, stand beside us, offering their support and understanding. The camaraderie that was forged in the crucible of service will continue to be a source of solace and inspiration as we navigate this new chapter of our lives.

So take a deep breath, for the winds of change are upon us. Embrace the challenges with determination, knowing that you possess the strength and resilience to overcome them. As we turn the page and begin this new chapter, let us embark on this transition journey together, knowing that we are capable of forging a bright and fulfilling future beyond the uniform.

INTRODUCTION

Transitioning out of the military is a significant life event that marks the end of a dedicated chapter in one's life and the beginning of a new journey. For service members, the shift from a highly structured and disciplined military environment to civilian life can be both exciting and challenging. It entails adapting to a world with different expectations, opportunities, and norms. This transition is not just about finding a new career; it involves redefining one's identity, establishing a sense of purpose, and integrating into civilian society.

The process of transitioning out of the military is multifaceted, requiring careful planning, introspection, and support. Each service member's experience is unique, shaped by their personal goals, skills, and circumstances. While some individuals may choose to pursue higher education, others may opt for vocational training, entrepreneurship, or joining the workforce directly. Regardless of the chosen path, transitioning requires a comprehensive approach that addresses various aspects, including education, employment, financial stability, and overall well-being.

One of the key challenges faced during this period is the adjustment to a less-regimented lifestyle. The military's structured routines, hierarchical systems, and shared camaraderie provide a sense of belonging and purpose that can be hard to replicate in civilian life. Service members may encounter difficulties in finding their new identity, reconnecting with family and friends, and establishing a work-life balance.

Additionally, translating military skills and experience into civilian terms can be a complex task. Employers often have a limited understanding of the military culture and may struggle to recognize the transferable skills acquired during military service. Service members need to effectively communicate their strengths, leadership abilities, adaptability, and problem-solving skills to employers in a way that resonates with the civilian job market.

Recognizing these challenges, governments, nonprofit organizations, and support networks have implemented numerous initiatives to aid in the transition process. Transition assistance programs, educational grants, career-counseling services, and networking opportunities are available to help service members navigate their way into civilian life successfully. These resources aim to provide guidance, mentorship, and practical support to ensure a smoother transition and maximize the chances of success in the civilian sector.

In this ever-evolving world, the skills, values, and experiences gained in the military are highly valuable and sought after in various industries. The discipline, resilience, teamwork, adaptability, and problem-solving abilities developed during military service can be advantageous assets in the civilian workplace. With proper guidance, preparation, and determination, transitioning service members can thrive and embark on fulfilling postmilitary careers, making significant contributions to society in new and meaningful ways.

As we delve deeper into the intricacies of transitioning out of the military, we will explore the essential steps, challenges, opportunities, and resources available to support service members in their journey toward a rewarding civilian life. Whether you are an active duty member considering your postservice options or a family member seeking guidance, this comprehensive book aims to equip you with the knowledge and tools necessary to navigate this transformative phase with confidence and success.

Transitioning out of the military is a significant life change that comes with its own unique set of challenges for veterans. While each individual's experience may vary, there are several common hurdles that many veterans encounter during this period. Understanding these challenges is crucial to providing the necessary support and resources for a successful transition. Here are some of the biggest challenges veterans face when transitioning out of the service:

1. Identity and Purpose: For many veterans, their military service becomes an integral part of their identity. Leaving behind the structured military environment can lead to a loss of identity and a sense of purpose. Finding a new purpose and establishing a civilian identity can be a complex and introspective process.
2. Civilian Culture Gap: The military has its own distinct culture, language, and values, which may differ significantly from the civilian world. Veterans may struggle to adapt to the different norms, behaviors, and expectations of civilian society. This cultural gap can manifest in various areas, including social interactions, workplace dynamics, and even family relationships.
3. Translating Military Skills: Translating military skills and experiences into civilian terms can be challenging. Employers may struggle to understand the relevance and applicability of military training, making it difficult for veterans to effectively communicate their abilities and experiences in the civilian job market. This disconnect can lead to underemployment or difficulties in securing suitable employment.
4. Limited Civilian Networks: The military provides a strong support system and a built-in community. Transitioning out of the service often means leaving behind the close-knit bonds and connections formed during military service. Veterans may find themselves lacking a similar support network in the civilian world, which can lead to feelings of isolation and loneliness.
5. Mental Health and Well-Being: The military exposes service members to unique stressors, including combat experiences, prolonged separations from loved ones, and constant vigilance. Transitioning out of the military can exacerbate mental health challenges such as post-traumatic stress disorder (PTSD), depression, anxiety, and adjustment disorders. Accessing appropriate mental health services and support becomes crucial during this period.

6. Financial Stability: Transitioning from a steady military paycheck to civilian employment can pose financial challenges. Veterans may need to navigate the complexities of civilian salaries, benefits, and retirement plans. Financial planning and management become critical to ensure a smooth transition and long-term stability.
7. Education and Training: Many veterans choose to pursue further education or training to enhance their civilian career prospects. However, navigating the education system, identifying suitable programs, and accessing educational benefits can be overwhelming. Balancing educational goals with other responsibilities and adjusting to a different learning environment can pose additional challenges.
8. Employment Opportunities: Veterans have a unique set of skills and experiences that make them highly desirable candidates in the job market, opening up a wide range of employment opportunities. Many employers actively seek out veterans due to their strong work ethic, discipline, leadership abilities, and problem-solving skills, creating a favorable landscape for veterans to secure meaningful employment.
9. Health Care and Benefits: Transitioning veterans often need to navigate the intricacies of the civilian health-care system and understand their eligibility for health-care benefits. Accessing quality health care, including specialized services for service-related injuries or conditions, can be a complex process.
10. Housing: When transitioning from military service to civilian life, veterans often encounter significant housing struggles. They may face difficulties in finding affordable and stable housing due to financial constraints, limited rental options, and the competitive housing market. Additionally, the process of adjusting to civilian life, coupled with potential mental health issues, can further complicate their housing situation, making the transition even more challenging.

Addressing these challenges requires a comprehensive and coordinated approach from government agencies, nonprofit organizations, employers, and communities. Transition assistance programs, career counseling services, mentorship initiatives, educational benefits, and mental health support can play a crucial role in easing the transition and empowering veterans to successfully integrate into civilian life. By recognizing and proactively addressing these challenges, we can honor the sacrifices and service of our veterans and help them achieve meaningful and fulfilling lives beyond the military.

Chapter 1

Identifying Purpose

Leaving the military marks a significant transition in a veteran's life, often bringing about a sense of uncertainty and the need to redefine one's purpose. In the military, service members are driven by a clear mission, a sense of duty, and a strong camaraderie. However, transitioning to civilian life can leave veterans searching for a new sense of purpose and direction. In this chapter, we explore the journey of identifying purpose after leaving the service, delving into the challenges, strategies, and resources available to veterans in this introspective quest.

1. The Importance of Purpose: Purpose is a driving force that gives life meaning and direction. It provides a sense of fulfillment, motivation, and a reason to get up every day. Identifying and embracing purpose is crucial for veterans transitioning out of the service as it contributes to their overall well-being, mental health, and successful reintegration into civilian life.

Purpose is a fundamental aspect of human existence that brings meaning, direction, and fulfillment to our lives. This significance is amplified for veterans transitioning out of the military, as they undergo a profound shift in their identity and daily routines. The pursuit and discovery of purpose play a crucial role in the successful reintegration of veterans into civilian life. A clear sense of purpose contributes significantly to a veteran's mental and emotional well-being. After leaving the service, veterans may experience a range of emotions, including loss, isolation, and a lack of direction. Purpose provides a solid foundation for psychological resilience, helping veterans navigate the challenges and uncertainties they encounter. It offers a reason to persevere, even in the face of adversity, fostering a sense of hope, motivation, and overall psychological well-being.

The military often becomes an integral part of a veteran's identity. Leaving behind that identity can lead to a loss of self and a sense of disorientation. Purpose acts as a compass, guiding veterans as they redefine their identity and establish a new sense of self-worth. It allows them to anchor their worth and value beyond their military service, recognizing their unique qualities, skills, and potential contributions to society.

Purpose offers veterans a distinct path and a profound purpose in their lives. It serves as a driving force that propels them forward, enabling them to set and pursue meaningful goals in civilian life. Having a purpose gives veterans a reason to wake up each day with determination, focus, and a sense of direction. It fuels their motivation, inspiring them to continue making a positive impact and contributing to their communities.

Transitioning out of the military is a complex process that requires adapting to a new environment, establishing new routines, and navigating unfamiliar systems. Purpose acts as an anchor during this period of change, providing a stable point of reference and guiding light. It supports veterans in their journey of reintegration by offering a sense of stability, clarity, and a sense of belonging in their new civilian lives.

Purpose cultivates a feeling of unity and companionship among veterans. When veterans find their purpose, they often gravitate toward like-minded individuals who share similar passions and goals. This creates a supportive network of individuals who understand and appreciate the unique experiences and challenges faced by veterans. The shared sense of purpose and connection within this network enhances social support, combats feelings of isolation, and strengthens the overall well-being of veterans.

Purpose empowers veterans to sustain their positive influence and make meaningful contributions to causes that surpass their selves. It provides them with a platform to utilize their unique skills, experiences, and perspectives acquired during their military service. Whether through civilian careers, volunteer work, mentoring, or advocacy, veterans can channel their purpose into actions that bring about positive change and improve the lives of others.

Living a purposeful life leads to long-term satisfaction and fulfillment. When veterans align their actions with their values and passions, they experience a deep sense of contentment and personal fulfillment. Purpose provides a sense of accomplishment and meaning, enhancing overall life satisfaction and promoting a positive outlook on the future.

For veterans transitioning out of the military, purpose carries profound significance. It serves as a vital component of their well-being, identity, and successful reintegration into civilian life. By identifying and embracing their purpose, veterans can forge a path that aligns with their values, passions, and unique contributions, ultimately leading to a fulfilling and meaningful postservice lifestyle.

2. Exploring Personal Passions and Interests: To identify purpose, veterans must first embark on a journey of self-discovery. This involves exploring personal passions, interests, and values. Reflecting on activities that bring joy, fulfillment, and a sense of accomplishment can unveil potential areas of purpose in a veteran's life. Engaging in hobbies, volunteering, and trying new experiences can help veterans reconnect with their authentic selves and discover new passions.

Exploring personal interests is a crucial endeavor for veterans as they transition out of the military and seek to redefine their identities and find fulfillment in civilian life. The process of discovering and engaging in activities that align with their passions and interests holds immense importance for veterans' overall well-being, personal growth, and successful postservice journey. Exploring personal interests allows veterans to embark on a journey of self-discovery. After years of dedicating themselves to military service, veterans may find that they have lost touch with their desires and interests. By engaging in activities that genuinely resonate with them, veterans have the opportunity to reconnect with their authentic selves, rediscover their passions, and gain a clearer sense of their identity beyond their military roles.

Engaging in activities that align with personal interests has a profound impact on veterans' emotional well-being. Pursuing hobbies and interests provides an outlet for stress reduction, relaxation, and enjoyment. It allows veterans to experience positive emotions, find joy in their daily lives, and counterbalance the challenges and stressors associated with the transition process. Exploring personal interests can help veterans foster a sense of contentment, balance, and overall emotional well-being.

Venturing into personal interests opens doors for veterans to pursue personal growth and development. By trying new activities and stepping outside their comfort zones, veterans expand their knowledge, skills, and perspectives. They acquire new experiences, gain confidence, and develop a sense of competence in areas outside their military expertise. This continuous growth contributes to their overall resilience, adaptability, and capacity to navigate diverse situations in civilian life.

Participating in personal interests allows veterans to connect with individuals who share similar passions and hobbies, fostering opportunities for meaningful connections. By participating in clubs, groups, or communities centered around their interests, veterans can establish new friendships and build support networks. These connections foster a sense of belonging, combat isolation, and provide a valuable source of social support during the transition process.

By exploring personal interests, individuals can cultivate and refine transferable skills and abilities that have broad applications, including in career endeavors. Veterans may discover hidden talents or develop skills that are directly relevant to their civilian careers or entrepreneurial endeavors. For instance, engaging in a hobby that requires teamwork and communication can enhance their interpersonal skills, which are highly valuable in the workplace.

Personal interests often reflect what brings veterans a sense of meaning and purpose. By exploring and engaging in activities they are genuinely passionate about, veterans tap into a source of intrinsic motivation and fulfillment. Personal interests provide a sense of purpose beyond professional or family obligations, allowing veterans to pursue activities that bring them joy, fulfillment, and a sense of accomplishment.

Engaging in personal interests provides veterans with healthy and constructive coping mechanisms. Immersing themselves in activities they enjoy serves as a form of relaxation, stress reduction, and a means of channeling their energy positively. It can help veterans manage transitional stress, improve their overall mental well-being, and maintain a balanced lifestyle.

As veterans transition into civilian life, the exploration of personal interests becomes highly significant. It facilitates self-discovery, emotional well-being, personal growth, and the development of valuable skills. By pursuing activities that align with their passions, veterans enhance their overall satisfaction, sense of purpose, and ability to adapt to the challenges of postservice life.

3. Assessing Skills, Talents, and Transferrable Qualities: Military service equips veterans with a vast array of skills and qualities that are highly valuable in civilian life. Identifying and leveraging these transferable skills is vital in finding purpose beyond the service. Veterans can evaluate their leadership abilities, teamwork skills, problem-solving capabilities, adaptability, and resilience to determine how these attributes can be applied in different professional and personal contexts.

Assessing skills, talents, and transferable qualities is a critical step for veterans as they transition from military service to civilian life. Veterans possess a unique set of abilities and experiences acquired during their military careers, and recognizing and leveraging these assets is essential for their successful integration into civilian society. Assessing skills, talents, and transferable qualities allows veterans to recognize the value and potential they bring to the civilian workforce and other areas of their lives. The military provides extensive training and development in areas such as leadership, teamwork, problem-solving, adaptability, and resilience. By identifying and acknowledging these skills, veterans can understand the breadth and depth of their capabilities, boosting their confidence and self-worth as they transition into civilian roles.

By evaluating their skills and transferable qualities, veterans can adeptly convey their abilities to prospective employers, educational institutions, and individuals. The challenge lies in translating military jargon, acronyms, and job titles into language that is easily understood and valued in the civilian world. By assessing their skills and qualities, veterans can articulate their experiences in a manner that highlights their relevance and applicability in various civilian contexts, increasing their chances of securing suitable employment or educational opportunities.

Through the evaluation of skills, talents, and transferable qualities, veterans gain the ability to make well-informed decisions regarding their career transitions. By identifying the skills they excel in and the areas they are passionate about, veterans can align their career goals with their strengths and interests. This assessment process empowers veterans to explore new career paths, upgrade their skills through training or education, or pursue entrepreneurial endeavors based on their transferable qualities, increasing their potential for professional advancement and long-term success.

The assessment of skills, talents, and transferable qualities presents veterans with avenues for personal development and growth. By identifying areas for improvement or skills they wish to further develop, veterans can seek out resources such as educational programs, vocational training, or mentorship to enhance their competencies. This commitment to lifelong learning fosters personal growth, adaptability, and a continuous improvement mindset, which are highly valued traits in the civilian workforce.

During the transition process, veterans can develop confidence and resilience by evaluating their skills, talents, and transferable qualities. Recognizing their capabilities and understanding the value they bring to different situations bolsters their self-assurance. This confidence serves as a foundation for resilience, empowering veterans to navigate challenges, setbacks, and rejections with a positive mindset and a belief in their ability to overcome obstacles.

By assessing their skills, talents, and transferable qualities, veterans can expand their horizons and gain a broader perspective on the diverse range of opportunities available to them. While some skills directly align with their military occupational specialties, veterans may discover other talents and qualities that can be applied in various fields. Assessing their skill set allows veterans to explore different industries, roles, and sectors where their abilities and experiences can make a significant impact, opening doors to new and fulfilling opportunities.

Veterans can effectively allocate their resources, such as time, energy, and effort, by assessing their skills, talents, and transferable qualities. By understanding their strengths and areas of expertise, veterans can focus on activities and opportunities that align with their skill set and interests. This targeted approach maximizes their potential for success and satisfaction, avoiding unnecessary diversions or pursuits that may not fully utilize their unique qualities.

4. Aligning Values and Beliefs: A sense of purpose often stems from aligning personal values and beliefs with one's actions and goals. Veterans can take the time to reflect on their core values and principles. Understanding what truly matters to them and how these values can guide their decisions and actions in civilian life can help shape a meaningful and purposeful path.

Aligning values and beliefs is a crucial aspect of the postservice journey for veterans transitioning into civilian life. The military instills in its members a strong sense of duty, honor, and service to a greater cause. As veterans enter a new phase of their lives, it becomes essential for them to assess and align their values and beliefs with their actions and goals. Veterans can lead authentic lives that align with their true selves by ensuring their values and beliefs are in harmony. It enables them to act with integrity, ensuring that their actions and decisions align with their core principles. By living by their values, veterans experience a sense of congruence between their internal beliefs and their external behaviors, promoting self-respect, personal satisfaction, and a sense of wholeness.

The alignment of values and beliefs plays a crucial role in discovering profound meaning and purpose in life. Veterans often have a deep sense of purpose ingrained in their military service, driven by a commitment to a greater cause. When they align their values and beliefs with their postservice goals, they can continue to pursue endeavors that resonate with their passions and sense of meaning. This alignment provides a sense of purpose beyond their military service, guiding their actions and giving direction to their postservice journey.

The positive alignment of values and beliefs has a beneficial impact on the overall well-being and mental health of veterans. When their actions align with their deeply held values and beliefs, they experience a sense of internal harmony and psychological well-being. On the other hand, a misalignment between personal values and actions can lead to cognitive dissonance, stress, and emotional turmoil. By aligning values and beliefs, veterans can cultivate a sense of fulfillment, contentment, and overall mental well-being.

Veterans can utilize the alignment of their values and beliefs as a framework for making decisions and establishing goals. When faced with choices or opportunities, veterans can evaluate them against their values and beliefs to determine whether they align with their authentic selves. This alignment serves as a compass, guiding veterans in making decisions that resonate with their true priorities and leading them toward a fulfilling and purposeful direction.

Aligning values and beliefs facilitates the formation of authentic relationships with others. By surrounding themselves with individuals who share similar values and beliefs, veterans can establish connections based on shared principles and mutual understanding. These relationships provide a support network and a sense of belonging, where veterans can be their authentic selves without compromising their values, fostering a positive social environment.

The alignment of values and beliefs equips veterans with an ethical and moral foundation to guide their actions. The military upholds values such as integrity, honor, and selflessness. By aligning their values with these foundational principles, veterans can continue to uphold high ethical standards and make a positive impact on their communities. This alignment reinforces their commitment to a life of integrity and service, both personally and professionally.

Aligning values and beliefs enhances veterans' resilience and adaptability as they navigate the challenges of postservice life. Values and beliefs act as guiding principles during times of uncertainty and change, providing a solid foundation and a source of stability. Veterans can rely on their deeply held values to guide their decisions, adapt to new environments, and overcome obstacles with resilience, maintaining their sense of self and purpose.

5. Engaging in Meaningful Work: Meaningful work plays a pivotal role in finding purpose. Veterans can explore career paths that align with their interests, values, and skills. This may involve pursuing a new profession, starting a business, or engaging in meaningful employment opportunities that contribute to a greater cause. Veterans can also consider avenues such as public service, advocacy, or nonprofit work that allow them to continue making a positive impact on society.

As veterans transition from military service to civilian life, the significance of engaging in meaningful work cannot be overstated. Meaningful work goes beyond simply earning a paycheck; it involves finding a sense of purpose, fulfillment, and personal satisfaction in one's professional endeavors. For veterans, meaningful work plays a vital role in their successful postservice journey by providing a platform to continue serving a greater cause, utilizing their unique skills, and making a positive impact on their communities.

Through active participation in meaningful work, veterans can uphold their sense of service and make valuable contributions to causes that extend beyond themselves. The military instills a strong sense of duty and service, and transitioning veterans often seek to find avenues where they can continue to make a positive impact. Meaningful work provides an opportunity to channel their skills, expertise, and values into professions that align with their desire to serve and contribute to society, fostering a sense of purpose and fulfillment.

Meaningful work helps veterans establish a sense of identity beyond their military service. Leaving the military can lead to a loss of identity and a sense of disorientation. Engaging in work that aligns with their values and passions allows veterans to redefine their identities and recognize the value they bring to their chosen profession. Meaningful work serves as a platform for veterans to showcase their unique skills, experiences, and talents, enhancing their self-esteem and overall sense of self-worth.

Engaging actively in meaningful work positively impacts the mental health and overall well-being of veterans. Meaningful work provides a sense of structure, purpose, and accomplishment, which are vital for psychological well-being. It gives veterans a reason to wake up each day with motivation and a sense of direction. By engaging in work that aligns with their values and passions, veterans experience a greater sense of satisfaction, fulfillment, and happiness, leading to improved mental health outcomes.

By actively participating in meaningful work, veterans can utilize and further develop the unique skills they acquired during their military service. Veterans possess a wide range of transferrable skills, such as leadership, teamwork, problem-solving, adaptability, and resilience. Meaningful work provides a platform to apply and further develop these skills in a civilian context. By utilizing their skills in a meaningful way, veterans enhance their professional competence, career prospects, and potential for long-term success.

Through engagement in meaningful work, veterans can cultivate a sense of belonging and integration within their communities. Meaningful work often involves collaboration with colleagues, clients, and community members. Through their work, veterans have the opportunity to establish connections, build relationships, and contribute to the betterment of their communities. Meaningful work creates a sense of belonging and social connectedness, which are crucial for veterans' successful integration into civilian life.

Veterans experience personal growth and fulfillment by actively participating in meaningful work. Continuous learning, skill development, and professional advancement are made possible through engaging in meaningful work. It challenges veterans to expand their knowledge, capabilities, and perspectives. By engaging in work that resonates with their passions and values, veterans experience a sense of personal fulfillment, accomplishment, and satisfaction, leading to a greater sense of overall well-being.

Through active involvement in meaningful work, veterans are allowed to make a positive difference in their communities and contribute to the betterment of society. Such work frequently entails contributing to the improvement of others' lives, whether through direct service, innovation, leadership, or advocacy. By engaging in work that aligns with their values, veterans can bring about positive change and improve the lives of those around them.

6. Embracing Lifelong Learning: Continuous growth and learning are key components of a purposeful life. Veterans can seek educational opportunities, professional development programs, and mentorship to expand their knowledge and skills. By embracing lifelong learning, veterans can evolve personally and professionally, opening doors to new possibilities and areas of purpose.

Embracing lifelong learning is of paramount importance for veterans as they transition from military service to civilian life and beyond. The military provides rigorous training and education, but the learning process does not end with the completion of service. Lifelong learning allows veterans to continuously acquire new knowledge, skills, and perspectives, empowering them to adapt to a rapidly changing world, pursue personal and professional growth, and thrive in their postservice journey. Veterans who embrace lifelong learning acquire the essential skills to thrive in an ever-changing world. The postservice landscape is constantly changing, whether in terms of technology, industry trends, or societal shifts. Lifelong learning enables veterans to stay current and relevant, ensuring their skills and knowledge remain up to date. By embracing continuous learning, veterans enhance their adaptability, resilience, and capacity to navigate new challenges and opportunities.

Continuing education and lifelong learning play a vital role in the career progression and professional growth of veterans. As they transition into civilian careers, veterans may need to acquire new skills or expand their knowledge to thrive in their chosen fields. By actively engaging in lifelong learning, veterans can enhance their qualifications, broaden their expertise, and improve their marketability. Lifelong learning provides access to additional certifications, degrees, and training programs, equipping veterans with the skills and knowledge necessary for career progression and new job opportunities.

Embracing continuous learning fosters personal growth and fulfillment among veterans. Learning new subjects, acquiring new skills, and pursuing intellectual curiosity contribute to a sense of self-actualization. Lifelong learning allows veterans to explore new interests, expand their horizons, and discover new passions. It stimulates intellectual engagement, critical thinking, and creativity, fostering a sense of purpose, personal fulfillment, and a zest for lifelong discovery.

The pursuit of lifelong learning enriches veterans' problem-solving and critical-thinking capabilities. The military instills a strong foundation of analytical thinking and problem-solving skills, but these skills can be further honed through ongoing learning. Lifelong learning exposes veterans to new ideas, perspectives, and approaches, challenging their assumptions and expanding their problem-solving toolkits. By embracing lifelong learning, veterans develop the ability to think critically, analyze complex situations, and make informed decisions in various aspects of their lives.

Veterans foster resilience and adaptability by embracing lifelong learning. Learning new skills and acquiring new knowledge requires a willingness to step out of one's comfort zone, embrace challenges, and persist in the face of obstacles. Lifelong learning encourages veterans to adopt a growth mindset, viewing failures and setbacks as opportunities for learning and growth. By cultivating resilience and adaptability through lifelong learning, veterans are better equipped to navigate the uncertainties and transitions they may encounter in their postservice journey.

Lifelong learning offers veterans opportunities to build social connections and participate in networking activities. Learning environments, whether in-person or online, bring together individuals with shared interests and goals. By actively participating in lifelong learning activities, veterans can connect with like-minded individuals, build relationships, and expand their professional and social networks. These connections can lead to mentorship opportunities, collaborative projects, and a sense of belonging within a community of learners.

Veterans can make valuable contributions to society by sharing their knowledge and expertise, thanks to lifelong learning. Veterans possess a wealth of experience and insights gained through their military service. By engaging in lifelong learning, veterans can transfer their knowledge to others as well as mentor and guide junior members of the community as they come into their role in society.

7. Building Supportive Networks: Navigating the journey of identifying purpose is not a solitary endeavor. Building supportive networks and connecting with fellow veterans, mentors, and community organizations can provide invaluable guidance, encouragement, and a sense of belonging. Veterans can seek out peer support groups, veteran-focused organizations, and mentoring programs that foster connections with like-minded individuals who share similar experiences.

Building supportive networks is of utmost importance for veterans as they transition from military service to civilian life. The military provides a strong sense of camaraderie, teamwork, and support, and veterans must establish similar networks in their civilian lives. Supportive networks provide a platform for veterans to connect with like-minded individuals, access resources, receive emotional support, and navigate the unique challenges they may face during their postservice journey. Establishing networks of support is crucial for promoting the emotional and mental well-being of veterans. Leaving the structured environment of the military can lead to feelings of isolation, disconnection, and loneliness. Supportive networks provide a sense of belonging and social connection, reducing the risk of mental health issues such as depression and anxiety. Connecting with fellow veterans or individuals who understand the military experience allows for shared experiences, empathy, and a safe space to discuss challenges and concerns.

Supportive networks play a crucial role in facilitating veterans' transition and integration into civilian life. Veterans often face unique challenges, such as translating their military skills to civilian job requirements, adjusting to different work cultures, or finding a sense of purpose and identity outside of the military. Supportive networks provide access to valuable resources, guidance, and mentorship, helping veterans navigate these challenges more effectively. They can offer advice on career opportunities, educational programs, and community services tailored to veterans' needs, easing the transition process and facilitating successful integration.

Creating supportive networks provides veterans with access to professional opportunities. Supportive networks often include individuals from various industries, organizations, and sectors. Connecting with professionals in these networks can lead to job referrals, mentorship, and access to hidden career opportunities. These networks can also provide insights into the job market, industry trends, and skill development needs, allowing veterans to stay informed and competitive in their desired fields.

Veterans greatly benefit from the invaluable peer support and empowerment offered by supportive networks. Interacting with fellow veterans who have faced similar challenges and experiences creates a sense of camaraderie and mutual understanding. Within these networks, veterans can share their triumphs, setbacks, and lessons learned, providing inspiration and encouragement to one another. This peer support instills a sense of empowerment, resilience, and motivation, reminding veterans that they are not alone in their journey and that they can overcome obstacles and achieve their goals.

For veterans, supportive networks serve as a pathway to beneficial resources and information. Through these networks, veterans can access organizations, programs, and services specifically designed to support their needs. These resources may include mental health services, educational opportunities, career counseling, financial assistance, and legal support. Building supportive networks allows veterans to tap into a wealth of knowledge and expertise, ensuring they have access to the support they require to thrive in their postservice lives.

Supportive networks provide a platform for veterans to advocate for their rights, raise awareness about issues affecting their community, and effect positive change. By uniting under a shared cause, veterans can amplify their voices and advocate for policies, programs, and services that address their unique needs. Supportive networks facilitate collaboration, mobilization, and collective action, empowering veterans to shape policies and initiatives that support their well-being and promote a smooth transition for future veterans.

Building supportive networks allows veterans to establish lifelong connections and friendships. The bonds formed within these networks are often deep rooted and enduring. Veterans can find solace, encouragement, and companionship among individuals who have shared similar experiences throughout their journeys and embrace each other in moments of need.

8. Seeking Professional Guidance: Veterans may benefit from seeking the guidance of career counselors, life coaches, or mentors who specialize in assisting transitioning service members. These professionals can provide personalized support, help veterans explore their options, and provide insights into potential paths aligned with their skills and passions.

Seeking professional guidance is of paramount importance for veterans as they navigate the complex process of transitioning from military service to civilian life. Professional guidance offers specialized support, expertise, and resources to help veterans overcome challenges, make informed decisions, and achieve their postservice goals. Whether it is in the areas of career planning, mental health, financial management, or educational opportunities, seeking professional guidance can significantly enhance veterans' transition experience and promote their overall well-being.

As veterans navigate the civilian job market, they need to seek professional guidance in career planning and employment. Transitioning from military service to a civilian career can be daunting, as veterans may need assistance in translating their military skills and experiences into marketable qualifications. Professional guidance provides access to career counseling, resume writing, interview preparation, and job-placement services tailored to veterans' unique needs. By seeking professional guidance, veterans can gain valuable insights into industry trends, job search strategies, and networking opportunities, increasing their chances of securing meaningful employment.

Veterans' overall well-being relies on the importance of seeking professional guidance for mental health. The transition from military service to civilian life can be emotionally challenging, and veterans may experience difficulties such as PTSD, anxiety, depression, or adjustment issues. Professional guidance, including mental health counseling and therapy, provides a safe space for veterans to address their emotional and psychological needs. Mental health professionals specializing in working with veterans can offer evidence-based therapies, coping strategies, and support systems to help veterans manage their mental health and improve their overall quality of life.

Veterans can broaden their knowledge, improve their skills, and enhance their career prospects by seeking professional guidance in pursuing educational opportunities, thus opening doors for them. Professionals in the field of education can assist veterans in navigating the complex landscape of educational programs, including college admissions, financial aid, and vocational training. They can help veterans identify the educational paths that align with their interests, goals, and available resources. Professional guidance ensures that veterans make informed decisions about their educational pursuits, leading to meaningful learning experiences and improved job prospects.

As veterans transition to civilian life and handle their financial responsibilities, seeking professional guidance in financial management is vital. Professionals specializing in financial planning and management can provide advice on budgeting, debt management, retirement planning, and investment strategies. They can help veterans create financial plans that align with their goals, ensure financial stability, and build a solid foundation for their future. Professional guidance in financial management empowers veterans to make informed decisions about their financial well-being, mitigate financial stressors, and achieve long-term financial security.

Veterans who may need assistance with diverse legal issues find it important to seek professional guidance in legal matters. Veterans may need legal support related to disability claims, benefits, housing, family law, or transitioning-related matters. Professionals in the legal field, including veterans' advocates and lawyers specializing in veterans' affairs, can provide guidance, representation, and support. They can navigate complex legal processes, ensure veterans' rights are protected, and advocate for their best interests. Seeking professional guidance in legal matters helps veterans navigate potential legal challenges and ensures that they receive the assistance they are entitled to.

By seeking professional guidance in entrepreneurship and accessing small business support, veterans can be empowered to pursue their entrepreneurial dreams and launch their businesses. Veterans often possess valuable skills, discipline, and leadership qualities that lend themselves to entrepreneurship. Professionals in business development and entrepreneurship can guide business planning, financing options, market research, and networking.

Identifying purposes plays a crucial role in bridging the civilian cultural gaps that veterans may encounter during their transition from military service to civilian life. Veterans bring with them a unique set of values, skills, and experiences that may differ from those prevalent in civilian culture. Understanding and identifying their purposes, both individually and as a collective, helps veterans navigate these cultural gaps, find their place in civilian society, and build meaningful connections with the civilian population.

The act of identifying purposes allows veterans to forge meaningful connections with the civilian population. By recognizing their values, goals, and aspirations, veterans can establish common ground with individuals who share similar interests or have aligned purposes. This shared sense of purpose forms the basis for building relationships, creating social connections, and bridging the cultural gap between the military and civilian worlds. It allows veterans to engage in conversations, activities, and collaborations that resonate with their missions, fostering a sense of belonging and connection with the civilian community.

Through the process of identifying purposes, veterans can actively shape perceptions and challenge existing stereotypes within the civilian culture. Veterans often face preconceived notions or misconceptions about their experiences, skills, or abilities when interacting with civilians. By clearly articulating their purposes and sharing their stories, veterans can challenge these stereotypes and provide a more nuanced understanding of their military service. Identifying purposes allows veterans to showcase their values, achievements, and contributions beyond their military roles, helping civilians see them as individuals with diverse talents and aspirations.

The act of identifying purposes facilitates the integration of veterans into civilian society and fosters collaborative efforts between veterans and civilians. When veterans are aware of their purposes and values, they can actively seek opportunities to contribute to their communities and engage in activities aligned with their goals. This proactive approach enables veterans to collaborate with civilians who share similar purposes, interests, or causes, transcending the cultural gap and fostering cooperation toward shared objectives. Through joint efforts and collaboration, veterans and civilians can work together to address societal challenges, promote mutual understanding, and strengthen community bonds.

The process of identifying purposes enriches cultural competence for both veterans and civilians alike. Cultural competence refers to the ability to effectively interact and engage with individuals from different cultural backgrounds. For veterans, identifying their purposes allows them to understand and appreciate civilian values, customs, and norms, promoting cultural sensitivity and adaptability. Veterans can actively seek to learn about civilian culture, listen to diverse perspectives, and adapt their behaviors and communication styles accordingly. Similarly, civilians can develop cultural competence by engaging with veterans, understanding their purposes, and appreciating their unique perspectives and contributions.

The act of identifying purposes generates opportunities for mutual learning and growth between veterans and civilians. Veterans bring a wealth of knowledge, skills, and experiences from their military service that can enrich civilian society. By sharing their purposes and insights, veterans can contribute to the collective growth and development of their communities. Simultaneously, civilians can share their perspectives, knowledge, and experiences, contributing to veterans' understanding of civilian culture. This reciprocal exchange of information and perspectives fosters mutual learning, empathy, and a deeper appreciation for each other's backgrounds, ultimately narrowing the cultural gaps.

Identifying purposes empowers veterans to become cultural ambassadors, promoting understanding and integration within civilian society. As veterans articulate their purposes and engage in civilian life, they serve as role models and ambassadors of military culture. By embracing their purposes and confidently navigating civilian cultural gaps, veterans inspire others to reach their full potential inside and outside a professional environment.

Chapter 2

Civilian Culture Gap

Transitioning from military service to civilian life presents unique challenges for veterans, including adjusting to civilian culture. Veterans often encounter cultural gaps that can affect their social interactions, communication styles, and overall sense of belonging in civilian society. In this chapter, we will explore the struggles veterans face with civilian culture gaps and the potential ways to address these challenges.

1. Communication and Language: One of the significant struggles veterans face when reintegrating into civilian culture is communication and language differences. Military jargon, acronyms, and direct communication styles may not be familiar to civilians. Veterans may find it challenging to express themselves or convey their experiences effectively, leading to misunderstandings or feelings of isolation. Bridging this gap requires veterans to adapt their communication style and learn to effectively convey their experiences and perspectives in civilian terms.

When veterans transition from military service to civilian life, they often encounter significant communication and language struggles due to cultural gaps. Military service involves a unique language and communication style that may not readily align with civilian norms. These challenges can affect veterans' ability to express themselves, build connections, and fully integrate into civilian society. The military employs a wide range of jargon and acronyms that are unfamiliar to civilians. Veterans may find it challenging to adjust their communication style, as they are accustomed to using military terminology that may be misunderstood or confusing to civilians. This can create barriers to effective communication and hinder veterans' ability to convey their experiences and perspectives in civilian settings. Overcoming this challenge requires veterans to learn to adapt their language and find common ground by using terminology and expressions that civilians can relate to.

One of the challenges faced by both civilians and veterans is the confusion that arises from translating different acronyms and jargon. Military service and civilian life have distinct terminologies, acronyms, and jargon that can create communication gaps and hinder understanding. The military relies heavily on acronyms and technical terminology to efficiently communicate complex information. However, civilians who are unfamiliar with military jargon often find themselves perplexed when engaging in conversations with veterans. The sheer volume of acronyms used in the military can overwhelm civilians, leading to misunderstandings and misinterpretations. To bridge this gap, veterans can make a conscious effort to explain the meaning behind acronyms when conversing with civilians, and civilians can actively seek clarification to enhance their understanding.

In addition to acronyms, military culture and various industries have their specific jargon that can be challenging to translate. Veterans may struggle to adapt their communication style, especially when using language that civilians are unfamiliar with. This can hinder effective communication and may create barriers in professional and social interactions. Both veterans and civilians can take steps to foster mutual understanding by actively engaging in conversations, asking for clarifications, and avoiding assumptions about shared knowledge or terminology.

The confusion resulting from different acronyms and jargon can lead to communication breakdowns and misunderstandings between civilians and veterans. Misinterpretations can occur when civilians mistakenly assign different meanings to military acronyms or when veterans assume that civilians possess knowledge of military terminology. These breakdowns can strain relationships and impede effective collaboration. Both parties need to remain open-minded, patient, and willing to engage in open dialogue to bridge these communication gaps.

Education and dialogue are key to overcoming the confusion caused by translating acronyms and jargon. Veterans can play an active role in explaining military terminology to civilians, providing context, and clarifying any misunderstandings that may arise. Civilians can also take the initiative in familiarizing themselves with common military acronyms and jargon through resources and training programs. By fostering a culture of understanding and open communication, both veterans and civilians can contribute to a more inclusive and collaborative society.

To enhance communication between civilians and veterans, it is essential to develop effective strategies that facilitate understanding. Veterans can learn to adapt their language by utilizing civilian terms whenever possible or by providing explanations and examples to help civilians grasp the meaning behind military jargon. Civilians can actively listen, ask questions, and seek clarification when encountering unfamiliar acronyms or jargon. Encouraging patience, empathy, and a willingness to learn can significantly reduce confusion and improve communication between both parties.

Technology can play a valuable role in bridging the gap between civilians and veterans in terms of understanding acronyms and jargon. Online platforms, forums, and social media groups can provide spaces for veterans to explain military terms and for civilians to seek clarification. Additionally, organizations and educational institutions can develop resources, glossaries, or training modules that facilitate the translation of military acronyms and jargon for civilians.

The confusion that arises from translating different acronyms and jargon can pose challenges for both civilians and veterans in their interactions. By fostering mutual understanding, engaging in open dialogue, and leveraging education and resources, we can bridge the gap between veterans and civilians.

Military communication often emphasizes directness, clarity, and brevity. In contrast, civilian communication tends to be more nuanced and may involve subtleties, social niceties, and indirectness. Veterans accustomed to being straightforward and concise in their communication may encounter difficulties in navigating the more complex dynamics of civilian conversations. They may inadvertently come across as too direct or blunt, which can lead to misunderstandings or strained relationships. Learning to interpret and navigate these subtleties while still maintaining authenticity is crucial for effective communication in civilian contexts.

Communication styles vary across cultures and contexts, and veterans often bring a unique communication style characterized by directness and bluntness due to their military background. However, this communication style can clash with the more nuanced and indirect communication norms prevalent in civilian society.

The military culture emphasizes a clear chain of command, efficiency, and directness in communication. Veterans are accustomed to straightforward and concise communication that leaves little room for ambiguity or misinterpretation. This directness serves a practical purpose in high-pressure military environments, where efficiency and clarity are crucial for mission success. However, this communication style may be perceived as overly blunt or confrontational in civilian settings.

Civilian communication often involves subtleties, social niceties, and indirect language to convey meaning and navigate social dynamics. The focus is not solely on the message itself but also on maintaining harmonious relationships and preserving individual sensitivities. This indirectness may manifest in using polite language, softening critiques, or relying on nonverbal cues. The directness and bluntness of veterans' communication styles can be jarring for civilians who are accustomed to these nuances.

When veterans communicate with their direct and blunt style, it can lead to misinterpretations and unintended offenses in civilian interactions. Civilians may perceive veterans' directness as rudeness, insensitivity, or a lack of tact. This misalignment in communication styles can strain relationships, hinder collaboration, and create a sense of social disconnect. Both veterans and civilians must recognize and appreciate these differences in communication norms to avoid misunderstandings.

Bridging the gap between veterans' directness and civilian communication norms requires awareness and adaptation from both parties. Veterans can make a conscious effort to understand and respect civilian communication norms by recognizing the value of nuance and indirectness. They can learn to soften their language, consider the impact of their words, and adjust their communication style to be more sensitive to others' emotions and expectations.

Civilians, on the other hand, can develop empathy and actively listen to veterans' direct communication style without immediately interpreting it as confrontational. Recognizing that veterans' directness is not intended as personal criticism but rather a product of their training and experiences can foster understanding and reduce misinterpretations. By listening actively and seeking clarification when necessary, civilians can bridge the communication gap and develop more effective communication strategies.

To facilitate effective communication between veterans and civilians, organizations, community groups, and educational institutions can provide communication training and mediation services. Such training can help veterans understand and adapt to civilian communication norms while also promoting awareness among civilians about veterans' communication styles and the underlying reasons behind their directness. Mediation services can facilitate dialogue, clarify misunderstandings, and promote understanding between veterans and civilians.

The directness and bluntness that veterans bring from their military background can present challenges when navigating civilian cultural gaps in communication. By fostering awareness, empathy, and active listening, both veterans and civilians can bridge this divide and develop effective communication strategies. Training programs, dialogue, and mediation services can further support the understanding of communication differences and promote smoother interactions. Ultimately, by recognizing and appreciating these differences, we can foster better understanding and collaboration between veterans and civilians, contributing to a more supportive environment.

Nonverbal cues and body language play a significant role in communication, but their interpretation can differ between military and civilian cultures. Military personnel often rely on specific nonverbal signals and gestures to convey messages efficiently. However, civilians may interpret these cues differently or may not be familiar with their significance. This can result in miscommunication or misinterpretation, potentially leading to strained interactions or misunderstandings. Veterans need to be aware of these differences and learn to adapt their nonverbal communication to align with civilian cultural expectations.

Nonverbal communication, including body language and gestures, plays a crucial role in human interaction. Veterans often possess distinct nonverbal cues acquired during their military service that may differ from civilian cultural norms. These differences can create misunderstandings and hinder effective communication between veterans and civilians. Here, we will explore the challenges veterans face in navigating nonverbal cues in civilian contexts and discuss strategies to bridge this gap and enhance communication.

The military relies heavily on nonverbal cues to convey information efficiently, maintain discipline, and ensure operational effectiveness. Veterans are trained to interpret and utilize specific gestures, facial expressions, and postures that hold specific meanings within the military context. However, these nonverbal cues may not align with civilian cultural norms, leading to potential confusion or misinterpretation.

Nonverbal communication cues can vary significantly across cultures. Different cultures have distinct norms regarding eye contact, personal space, hand gestures, and other nonverbal expressions. Veterans, conditioned by military culture, may unintentionally display nonverbal cues that are inconsistent with civilian cultural expectations. This can result in misunderstandings, discomfort, or a sense of social disconnection between veterans and civilians.

Body language is a powerful means of communication, but its interpretation can vary between military and civilian contexts. For example, veterans may exhibit a more rigid and upright posture due to their military training, while civilians may adopt a more relaxed or casual stance. This mismatch in body language can create perceptions of distance, formality, or even intimidation, impacting the quality of interpersonal relationships.

Military training often emphasizes discipline and control over emotional expression, particularly during stressful situations. Veterans may exhibit a more restrained or stoic facial expression compared to civilians, who may display a wider range of emotions openly. This contrast can lead to misunderstandings or a perception that veterans are unresponsive or unemotional. Recognizing and understanding these differences in emotional display is essential for fostering empathy and effective communication.

Both veterans and civilians can play a role in bridging nonverbal communication gaps. Veterans can be mindful of their nonverbal cues and adapt them to align with civilian cultural norms. This includes adjusting their body language, facial expressions, and emotional display to be more in line with civilian expectations. Similarly, civilians can demonstrate openness, flexibility, and curiosity when encountering nonverbal cues that differ from their cultural background.

Enhancing cultural sensitivity and awareness is crucial for improving communication between veterans and civilians. Educational institutions, organizations, and community groups can provide cultural sensitivity training to help civilians understand and appreciate the nonverbal cues veterans may display. Veterans can also contribute by sharing their experiences and offering insights into the nonverbal cues they have learned during their military service.

Active listening and clarification play vital roles in bridging nonverbal communication gaps. Veterans and civilians should strive to listen attentively, seeking clarification when necessary to ensure an accurate understanding of nonverbal cues. Clear communication channels and open dialogue allow for the exchange of perspectives, leading to a greater appreciation of each other's nonverbal communication styles.

Nonverbal communication cues can significantly impact interpersonal interactions between veterans and civilians. By recognizing the differences in nonverbal communication and fostering cultural sensitivity and awareness, we can bridge the gap and promote effective communication. Both veterans and civilians can adapt their nonverbal cues, practice active listening, and seek clarification when needed to enhance understanding and effective communication to foster and develop a closer understanding of each other.

Military communication is hierarchical and often involves clear chains of command and direct instructions. In contrast, civilian communication tends to be more egalitarian, emphasizing collaboration, consensus-building, and participatory decision-making. Veterans may find it challenging to adapt to the more collaborative and inclusive communication styles prevalent in civilian workplaces or social settings. Recognizing and embracing these differences can help veterans navigate civilian communication dynamics more effectively and build positive relationships.

Communication styles vary among individuals, shaped by factors such as cultural background, personal experiences, and professional training. Veterans bring unique communication styles to interpersonal interactions due to their military service. Understanding and appreciating these different communication approaches is crucial for effective communication and building meaningful connections. Let's dive into the different communication styles that veterans may utilize and the importance of acknowledging and adjusting to these variations.

Military training often instills a communication style that is direct, assertive, and focused on clear instructions and objectives. Veterans are accustomed to concise and unambiguous language, emphasizing efficiency and precision. This communication style aims to minimize misunderstandings and promote effective teamwork. However, in civilian contexts, this directness can be perceived as overly blunt or aggressive. Recognizing the veterans' preference for direct communication can help civilians adapt their communication style and avoid misinterpretation.

Veterans often possess a mission-oriented mindset, focusing on tasks, goals, and objectives. Their communication style may reflect this priority, with an emphasis on efficiency and effectiveness. Veterans may appreciate brevity and straightforwardness in communication, striving to convey information quickly and directly. However, this mission-oriented approach may inadvertently overlook relational aspects or nuances present in civilian communication styles, leading to potential miscommunication or disconnect.

Military service requires adaptability and flexibility, skills that veterans bring to their communication styles. Veterans are trained to quickly adjust their communication approach based on the context, audience, and objectives. This ability to adapt allows them to effectively communicate in various settings and with individuals from diverse backgrounds. Recognizing and appreciating this adaptability can enhance collaboration and open doors for constructive dialogue between veterans and civilians.

Effective communication is a two-way process that involves active listening and providing feedback. Veterans are trained in active listening skills, which include focusing attention, paraphrasing, and seeking clarification to ensure accurate comprehension. This active listening approach promotes effective understanding and mutual respect. Encouraging and practicing active listening as civilians can foster better communication and demonstrate respect for the veterans' communication style.

Communication styles are deeply connected to building rapport and trust. Veterans often prioritize trust and loyalty due to their military experience, and their communication style may reflect this value. Understanding this aspect of veterans' communication approach can help civilians create a supportive environment where veterans feel valued and respected. By fostering open and nonjudgmental communication, civilians can build trust and strengthen relationships with veterans.

Education and awareness play vital roles in bridging communication gaps between veterans and civilians. Organizations, institutions, and communities can provide training programs or workshops to educate civilians about veterans' communication styles, emphasizing cultural sensitivity and understanding. Similarly, veterans can participate in programs that guide adapting their communication style to civilian contexts, promoting effective interaction and integration.

Different communication styles exist among veterans, shaped by their military background and experiences. By recognizing and appreciating these diverse communication approaches, both veterans and civilians can foster effective communication, build meaningful connections, and bridge the gap that may exist. Adapting communication styles, practicing active listening, and promoting education and awareness are key steps in effective communication and play a vital role in a veteran's transition process.

Veterans often struggle to articulate their military experiences and translate them into language that civilians can comprehend. The unique nature of military service, including combat experiences or highly specialized roles, may be difficult for civilians to fully understand or relate to. Veterans may feel frustrated or isolated when trying to share their stories or convey the significance of their experiences. Encouraging empathy, active listening, and providing platforms for veterans to share their stories can bridge this communication gap and foster mutual understanding.

The military experience is a profoundly transformative journey that shapes the lives of those who serve. Veterans often face the challenge of articulating their unique and complex experiences to others who have not shared the same journey. Expressing the military experience goes beyond merely recounting events; it involves conveying the emotional, psychological, and personal impact that service has had on their lives. We will examine the importance of veterans articulating their military experience and explore the various channels through which they can effectively convey the profound nature of their service.

Veterans have made significant sacrifices during their service, whether it be time away from loved ones, physical injuries, or the emotional toll of combat. Expressing the military experience allows veterans to communicate the depth of these sacrifices, providing others with a glimpse into the challenges they have faced and the resilience they have demonstrated. Through storytelling and personal narratives, veterans can convey the sacrifices they have made and the dedication they have shown to their country.

Military service often imparts valuable life lessons and personal growth. Veterans possess unique perspectives and insights derived from their training, experiences, and interactions with diverse individuals. By expressing their military experience, veterans can share these lessons learned with others, offering wisdom and knowledge that extends beyond the confines of the military. This sharing of experiences can foster mutual understanding, promote personal growth, and contribute to a broader societal appreciation of the military's contributions.

Veterans' stories and experiences have the power to inspire and motivate others. By expressing their military journey, veterans can serve as role models, demonstrating resilience, courage, and a commitment to service. These narratives can have a profound impact on individuals facing their challenges, offering hope, encouragement, and a sense of shared purpose. Veterans' expressions of their military experience can ignite a spark in others and contribute to a more empathetic and engaged society.

The military experience can often create a divide between veterans and civilians due to differences in perspectives, values, and lived experiences. Expressing their military experience allows veterans to bridge this gap and promote understanding. By sharing their stories, veterans can provide insights into the complexities of military service, dispel misconceptions, and foster empathy. This dialogue between veterans and civilians helps create a more inclusive society that appreciates the sacrifices and contributions of those who have served.

Expressing the military experience can take many forms, allowing veterans to find avenues that resonate with their unique voices and talents. Some veterans may choose to write memoirs or poetry, while others may express themselves through visual arts, music, or public speaking. Utilizing various mediums provides veterans with a range of creative outlets through which they can effectively communicate their military experience and engage with diverse audiences.

To encourage veterans to express their military experience, it is essential to create supportive spaces that promote active listening, respect, and understanding. Communities, organizations, and educational institutions can provide platforms for veterans to share their stories and experiences in a safe and nonjudgmental environment. These spaces foster a sense of belonging and validation for veterans, allowing them to express themselves authentically and share their journeys with receptive audiences.

Expressing the military experience is a significant endeavor that enables veterans to convey the depth of their service, share valuable insights, inspire others, and bridge the divide between veterans and civilians. Through storytelling, artistic expression, and the creation of supportive spaces, veterans can articulate the impact of their military experience and contribute to a more informed and empathetic society. It is through these expressions that veterans find their voices and their stories become a vital part of our collective narrative that we all share.

Addressing the communication and language struggles that veterans face requires a collective effort from both veterans and civilians. Education and awareness initiatives can play a crucial role in bridging the cultural divide. Providing cultural sensitivity training to civilians can help them better understand and appreciate the unique communication challenges faced by veterans. Similarly, veterans can actively seek opportunities to improve their communication skills, such as enrolling in public-speaking courses or participating in workshops that focus on effective communication in civilian contexts.

Veterans bring a wealth of knowledge, experiences, and cultural perspectives gained through their military service. As they transition back into civilian life, veterans have the unique opportunity to bridge cultural gaps and foster understanding between themselves and the civilian community. Through education and awareness initiatives, veterans can share their insights, promote empathy, and contribute to a more inclusive society. This focus will be on the vital role that veterans can play in promoting mutual understanding and appreciation among civilians by bridging cultural gaps through education and increased awareness.

Military culture is characterized by a strong sense of duty, discipline, and camaraderie. Veterans can educate civilians about the core values instilled in them during their service, such as integrity, loyalty, and selflessness. By sharing these values and the importance they hold in military culture, veterans can promote a deeper understanding of the military mindset and bridge the cultural gap between veterans and civilians.

Misconceptions and stereotypes about veterans often contribute to misunderstandings and cultural gaps. Veterans can take an active role in dispelling these misconceptions through education and awareness initiatives. By sharing their personal stories and experiences, veterans can challenge preconceived notions and provide nuanced perspectives that counter stereotypes. This openness and willingness to engage in dialogue help civilians gain a more accurate understanding of veterans' lives and experiences.

Veterans can contribute to educational programs and training sessions aimed at promoting cultural sensitivity and awareness. Their firsthand knowledge of diverse cultures, gained through their military service, allows them to provide insights into different ethnicities, religions, and customs. By participating in these training initiatives, veterans can help civilians develop a deeper appreciation for cultural diversity, fostering an inclusive environment that respects and celebrates differences.

Veterans can play a crucial role in facilitating dialogue and discussion between themselves and civilians. By organizing forums, panels, or community events, veterans can create opportunities for open and honest conversations about their military experiences and the cultural gaps that exist. These platforms encourage active listening, empathy, and the exchange of perspectives, leading to mutual understanding and the bridging of cultural divides.

Veterans can serve as mentors and provide peer support to both fellow veterans and civilians. Their unique experiences and perspectives position them well to guide individuals through the challenges of transitioning to civilian life. Veterans can share insights, offer guidance, and provide a supportive network that helps individuals navigate the cultural gaps that may arise during this transition. This mentorship and peer support contribute to a more seamless integration of veterans into civilian communities.

Engaging with community organizations provides veterans with opportunities to make a significant impact in bridging cultural gaps. Veterans can collaborate with local nonprofits, schools, or diversity initiatives to develop programs that promote understanding and appreciation of military culture. By actively participating in these initiatives, veterans can contribute to a more inclusive society that recognizes and values the unique contributions of veterans.

Veterans can leverage their knowledge and experiences to advocate for policies and initiatives that support bridging civilian cultural gaps. By engaging with policymakers, community leaders, and advocacy groups, veterans can help shape policies that promote cultural sensitivity, inclusion, and support for veterans. Their insights and firsthand experiences provide a valuable perspective that can drive positive change at a systemic level.

Veterans have a vital role to play in bridging civilian cultural gaps through education and awareness. By sharing their military culture, dispelling misconceptions, promoting dialogue, and advocating for policy changes, veterans can foster understanding, empathy, and appreciation between veterans and civilians.

2. Differences in Values and Priorities: Military culture emphasizes discipline, hierarchy, and a mission-oriented mindset, which can differ significantly from civilian values and priorities. Veterans may find it challenging to navigate the civilian world, where individualism, personal pursuits, and different value systems prevail. Adjusting to civilian culture often involves reassessing personal values, understanding different perspectives, and finding common ground to build meaningful relationships and connections.

Veterans bring a distinct set of values and priorities to their civilian lives, shaped by their military experiences. However, the presence of cultural gaps between veterans and civilians can significantly influence how veterans' values and priorities are understood and appreciated in society. A fundamental value instilled in veterans is a strong sense of duty and service to their country. This value is deeply rooted in military culture, where the mission and the welfare of comrades take precedence. However, civilian cultural gaps can sometimes lead to a lack of understanding and appreciation for this sense of duty. Veterans may encounter challenges in aligning their values with societal norms and may struggle to find opportunities to continue serving their communities. Recognizing and valuing veterans' sense of duty can help bridge the cultural gap and create a more supportive environment for their continued service.

In the military, veterans are guided by a clear sense of duty, defined by their roles, missions, and responsibilities. However, as they transition to civilian life, their focus and responsibilities often undergo significant changes. Veterans may find themselves grappling with new questions about their purpose and how to contribute meaningfully to their communities. Navigating this shift requires introspection, exploration, and the recognition that their sense of duty can manifest in diverse ways outside the military context.

While the nature of service changes after leaving the military, many transitioning veterans still possess a strong desire to contribute to their communities. They seek opportunities to channel their skills, experiences, and sense of duty into meaningful service. Veterans may engage in volunteer work; join organizations focused on social causes; or pursue careers in public service, health care, or education. Recognizing and facilitating these avenues for service can empower veterans to continue making a positive impact in their postmilitary lives.

The transitioning veteran's sense of duty often extends beyond their professional responsibilities. They may face the challenge of balancing personal obligations, such as family, with their commitment to service. Veterans must navigate the demands of their careers while nurturing relationships, maintaining a healthy work-life balance, and meeting the expectations they set for themselves. Supporting transitioning veterans in achieving this balance is crucial for their well-being and successful integration into civilian life.

In the military, success is often measured by rank, promotions, and mission accomplishments. However, the transitioning veteran's sense of duty requires them to redefine success and achievement in civilian terms. Veterans may need support in recognizing and valuing their individual accomplishments, skills, and personal growth outside the military context. Encouraging a broader understanding of success can help veterans navigate their evolving sense of duty and find fulfillment in their postservice endeavors.

Transitioning veterans often possess a growth mindset and a willingness to embrace new challenges and opportunities. They understand the importance of continuous learning and personal development. To support their evolving sense of duty, it is crucial to provide access to educational resources, training programs, and mentorship opportunities. Facilitating their pursuit of knowledge and skills equips transitioning veterans with the tools they need to contribute meaningfully to their chosen fields and communities.

Building supportive networks is essential for transitioning veterans as they navigate their changing sense of duty. Peer support groups, mentorship programs, and community organizations can provide a sense of belonging, understanding, and guidance. These networks offer a space for veterans to share their experiences, exchange insights, and receive support from others who have undergone similar transitions. By fostering these networks, we create an environment that empowers transitioning veterans to continue embodying their sense of duty in new and meaningful ways.

The transitioning veteran's sense of duty undergoes a profound transformation as they navigate the complexities of civilian life. By recognizing the shifting focus and responsibilities, providing opportunities for service, supporting work-life balance, redefining success, promoting continuous growth, and fostering supportive networks, we can empower transitioning veterans to embrace their evolving sense of duty.

Veterans often make significant sacrifices during their military service, whether it be time away from loved ones, physical and mental challenges, or even the ultimate sacrifice of losing comrades. These sacrifices reflect the value of selflessness and devotion to a cause greater than oneself. However, civilian cultural gaps can lead to a lack of awareness and understanding of these sacrifices. Veterans may feel their sacrifices go unrecognized or undervalued, which can affect their sense of purpose and identity. Acknowledging and honoring the sacrifices made by veterans is crucial in bridging the cultural gap and demonstrating appreciation for their selflessness.

Veterans embody a deep sense of sacrifice and selflessness developed during their military service. This willingness to put others before themselves and make personal sacrifices shapes their identity and values. However, during the transition to civilian life, veterans face unique challenges in reconciling their sacrifices and selflessness with the new realities they encounter. Here we dive into how veterans handle their sacrifices and selflessness during the transition, highlighting the importance of understanding, support, and self-care throughout this journey.

The transition period offers veterans an opportunity to reflect on their sacrifices and selflessness during their military service. Veterans must acknowledge and validate their experiences, recognizing the significance of their sacrifices and the impact they had on their lives.

This reflection helps veterans develop a deeper understanding of their values and priorities as they navigate the transition to civilian life.

Veterans undergoing the transition process often encounter unfamiliar roles and responsibilities, both in their personal and professional lives. In the military, their sense of selflessness was deeply ingrained, with a focus on the mission and the welfare of their comrades. As they transition, veterans may need to redefine their understanding of selflessness to include personal aspirations, family commitments, and contributions to the civilian community. This adjustment requires a balance between honoring their previous selflessness and recognizing the need for self-care and personal growth.

One of the key challenges during the transition is finding purpose and meaning outside of the military context. Veterans may initially struggle to identify how their selflessness and sacrifices can be channeled into civilian life. Engaging in meaningful activities, such as volunteer work, mentorship, or pursuing a career that aligns with their values, can help veterans rediscover a sense of purpose and maintain their commitment to making a positive impact on the world.

Veterans often prioritize the well-being of others, sometimes at the expense of their self-care. During the transition, veterans need to recognize the importance of self-care and prioritize their own mental, emotional, and physical wellness. This may involve seeking professional help, participating in wellness programs, engaging in hobbies and activities that bring joy, and fostering healthy relationships. By taking care of themselves, veterans can better navigate the challenges of the transition period and continue to serve as positive role models.

Practicing gratitude and reflection is instrumental in helping veterans process their sacrifices and selflessness. By focusing on the positive aspects of their experiences, veterans can cultivate a sense of pride, resilience, and personal growth. Expressing gratitude for the support they receive during the transition and recognizing the opportunities that lie ahead allow veterans to navigate the complexities of their sacrifices with a sense of optimism and purpose.

As veterans progress in their transition journey, they can become advocates and mentors for other transitioning veterans. Sharing their own experiences, offering guidance, and providing support can help veterans reinforce their sense of selflessness by giving back to their community. By being a mentor and advocates, veterans can make a difference in the lives of others, embodying their continued commitment to service.

Military service fosters a strong sense of teamwork, camaraderie, and trust among veterans. The bonds formed in the military are often lifelong and deeply cherished. However, civilian cultural gaps may lead to a lack of understanding about the depth and significance of these connections. Veterans may struggle to find similar levels of camaraderie and support in civilian life, which can impact their overall well-being and sense of belonging. Recognizing the value of teamwork and fostering environments that promote connection and support can help bridge this cultural gap and support veterans' transition to civilian life.

Teamwork and camaraderie form the bedrock of military service, fostering deep connections and a shared sense of purpose among veterans. However, as veterans transition out of the service and enter civilian life, they often encounter challenges in replicating the same level of teamwork and camaraderie they experienced in the military.

One of the primary challenges veterans face during the transition is the loss of the built-in support systems inherent in military life. In the military, teamwork and camaraderie are cultivated through shared experiences, rigorous training, and a common mission. Transitioning out of the service can lead to a sense of isolation and a longing for the close-knit community veterans once had. Recognizing this loss and actively seeking ways to build new support networks are crucial for veterans' successful integration into civilian life.

To recreate the sense of camaraderie experienced in the military, veterans can actively seek out like-minded individuals who share their values, experiences, and aspirations. This can involve joining veteran support groups, participating in community organizations focused on veterans' issues, or engaging in hobbies and activities that attract individuals with similar backgrounds. By connecting with others who understand the unique challenges and experiences of military service, veterans can forge new bonds and build a supportive network.

Mentorship and peer support play vital roles in fostering teamwork and camaraderie during the transition. Veterans who have successfully navigated the challenges of reintegrating into civilian life can offer guidance, share insights, and provide a listening ear to those in similar situations. Mentorship programs and peer support groups create spaces for veterans to exchange experiences, offer advice, and provide emotional support. These relationships foster a sense of camaraderie and contribute to veterans' overall well-being.

Engaging in volunteer work and community initiatives provides veterans with opportunities to collaborate with others toward a common goal. By participating in projects that align with their values and interests, veterans can tap into their innate sense of teamwork and camaraderie. Serving alongside fellow community members fosters a sense of purpose and belonging, replicating the bonds veterans formed during their military service.

Employers play a critical role in fostering teamwork and camaraderie among veterans in the civilian workforce. Creating a supportive workplace culture that recognizes and values veterans' experiences, skills, and contributions is essential. Employers can implement transition programs that facilitate networking, mentorship, and team-building activities specifically tailored to veterans. These initiatives help veterans feel integrated within the workplace and enable them to leverage their team-oriented skills.

Nurturing teamwork and camaraderie is vital for veterans as they transition beyond the service. By acknowledging the loss of built-in support systems, translating team skills to civilian settings, connecting with like-minded individuals, engaging in mentorship and peer support, volunteering, and fostering supportive workplace cultures, veterans can build a successful network emphasized by the camaraderie from their past.

Military service cultivates resilience and adaptability in veterans, enabling them to navigate challenging and rapidly changing environments. However, civilian cultural gaps may not fully appreciate or recognize these qualities in veterans. Veterans may encounter difficulties in translating their resilience and adaptability into civilian contexts, such as the job market or interpersonal relationships. Acknowledging and valuing veterans' resilience can help bridge the cultural gap and create opportunities for their successful integration into civilian life.

Resilience and adaptability are integral traits developed through military service, enabling veterans to overcome challenges and thrive in demanding environments. However, as veterans navigate the transition to civilian life, they encounter cultural gaps that require them to apply their resiliency and adaptability in new ways.

The transition from military to civilian life often exposes veterans to cultural gaps that can hinder effective communication and understanding. These gaps may manifest in different ways, such as differences in language, values, social norms, and expectations. Veterans may face challenges in relating to civilian peers, expressing their experiences, and finding a sense of belonging in a culture that may not fully comprehend the military context. These cultural gaps require veterans to draw upon their resiliency and adaptability to navigate these new environments.

Resilience is a fundamental attribute developed during military service that enables veterans to withstand and overcome adversity. When faced with cultural gaps, veterans draw upon their resilience to adapt, learn, and grow. They approach unfamiliar situations with determination and resourcefulness, finding ways to bridge the divide and forge connections. The ability to bounce back from setbacks and remain focused on their goals empowers veterans to persevere in the face of cultural challenges.

Adaptability is another critical trait veterans possess, allowing them to adjust to new circumstances and environments. Veterans are accustomed to rapid changes, shifting priorities, and diverse teams, which cultivates their flexibility and openness to new experiences. This adaptability proves invaluable as veterans navigate civilian cultural gaps. They can embrace differences, modify their communication styles, and seek opportunities for learning and growth. By adapting to new cultural norms, veterans can foster understanding and bridge gaps in perception and communication.

Veterans actively seek cultural understanding as a means to bridge gaps between military and civilian cultures. They recognize the importance of educating themselves about civilian values, social dynamics, and expectations. Veterans may engage in dialogue, attend cultural competency training, and seek mentorship from individuals familiar with both military and civilian contexts. This dedication to understanding allows veterans to navigate cultural gaps with empathy and sensitivity, ultimately fostering stronger connections with civilian counterparts.

Veterans' resiliency and adaptability extend to their approach to building networks and partnerships to bridge cultural gaps. They actively seek opportunities to collaborate with civilian individuals and organizations, leveraging their strengths and unique perspectives. By forming diverse partnerships, veterans foster an environment of mutual respect and understanding, creating bridges between military and civilian cultures. These networks become invaluable sources of support, guidance, and cultural exchange.

Recognizing the need for broader societal awareness and understanding, veterans often become advocates for bridging cultural gaps between the military and civilian communities. They engage in educational initiatives, public speaking, and outreach efforts to promote dialogue, dispel misconceptions, and foster a more inclusive society. By sharing their stories and experiences, veterans contribute to a greater understanding of military culture, promoting empathy, and closing the cultural gaps that exist.

The journey of transitioning veterans through cultural gaps brings about personal growth and transformation. Veterans' resiliency and adaptability allow them to navigate these gaps while embracing new perspectives and broadening their horizons. As they bridge cultural divides, veterans develop a deeper understanding of themselves and others, cultivating resilience and adaptability in new contexts. This personal growth not only benefits veterans but also enhances the cultural fluency between civilians and veterans.

Military service operates within a structured and disciplined framework that shapes veterans' values and priorities. This structure provides a clear sense of purpose, expectations, and direction. However, civilian cultural gaps can sometimes perceive veterans' appreciation for structure and discipline as rigidity or inflexibility. Veterans may face challenges in adapting to more fluid and less hierarchical civilian environments. Recognizing and respecting veterans' appreciation for structure while offering support and guidance in navigating civilian settings can help bridge the cultural gap and facilitate their successful transition.

Structure and discipline are fundamental elements of military life that profoundly shape the experiences and behaviors of veterans. As veterans transition to civilian life, these qualities play a significant role in helping them navigate and close the cultural gaps that exist between military and civilian contexts.

One of the significant challenges veterans face in civilian cultural contexts is the communication gap. The military operates with a distinct language, acronyms, and jargon that can be foreign to civilians. Veterans' training in clear and concise communication, as well as their adherence to a hierarchical chain of command, may create barriers when interacting with civilians. However, veterans' ingrained structure and discipline provide them with a foundation for adapting their communication style and bridging this gap. By consciously adjusting their language, avoiding military jargon, and expressing ideas in a relatable manner, veterans can enhance understanding and establish effective communication with civilians.

The military instills a set of cultural norms and expectations that differ from those in civilian life. Veterans are accustomed to a highly structured environment with strict rules, codes of conduct, and a clear chain of command. In contrast, civilian culture often values individualism, flexibility, and autonomy. Adjusting to these cultural differences can be challenging for veterans, as their disciplined mindset may clash with the more fluid nature of civilian settings. However, veterans' inherent understanding and appreciation of structure and discipline can facilitate their adaptation to civilian cultural norms. By recognizing and respecting the cultural context, veterans can align their behaviors and actions accordingly, fostering mutual understanding and acceptance.

The incorporation of structure and discipline is crucial in veterans' management of time and pursuit of goals. In the military, adherence to schedules and deadlines is crucial for mission success. Veterans are adept at planning, organizing, and prioritizing tasks to achieve objectives efficiently. As they transition to civilian life, veterans can apply these skills to close cultural gaps. By managing their time effectively, setting clear goals, and demonstrating reliability, veterans exhibit a disciplined approach that promotes trust and cooperation with civilians. Their ability to follow through on commitments and meet expectations helps build bridges and overcome cultural differences.

While structure and discipline are associated with rigidity, veterans also possess the adaptability and flexibility necessary to navigate civilian cultural gaps. Military service often requires veterans to adapt quickly to changing circumstances, diverse environments, and unexpected challenges. This adaptability enables veterans to embrace new cultural norms, practices, and perspectives. Their disciplined approach allows them to assess the requirements of the civilian context and make adjustments while staying true to their values and principles. By demonstrating flexibility, veterans can bridge cultural gaps, cultivate empathy, and foster positive relationships with civilians.

Veterans' experience in a structured and disciplined environment positions them as potential mentors and leaders in civilian settings. Veterans can offer guidance, share their knowledge, and serve as role models for both fellow veterans and civilians. Their ability to provide structure and direction can help others understand and appreciate the value of discipline, particularly in bridging cultural gaps. By showcasing effective leadership and mentorship, veterans contribute to a more cohesive and integrated society where the positive aspects of military structure and discipline are embraced and understood.

3. Lack of Understanding or Misconceptions: Civilian society may have limited knowledge or misconceptions about military service and the experiences of veterans. This lack of understanding can lead to stereotypes, biases, or unrealistic expectations. Veterans may feel misunderstood or face insensitive questions, making it difficult to relate to civilians or establish meaningful connections. Raising awareness about military service, educating civilians about veterans' experiences, and promoting dialogue can help bridge the gap and foster mutual understanding.

A lack of understanding is a significant barrier that transitioning veterans often face when attempting to bridge civilian cultural gaps. As veterans transition from the military to civilian life, they encounter a different set of values, norms, and expectations that may be unfamiliar to them.

One of the primary factors contributing to the lack of understanding between veterans and civilians is the presence of cultural differences and misconceptions. Each group has distinct experiences, beliefs, and ways of life that shape their perspectives. Veterans may struggle to comprehend civilian values of individualism, diversity, and flexibility, while civilians may have limited knowledge of the military culture, hierarchy, and sacrifices. These differences can lead to misinterpretations, stereotypes, and misunderstandings, creating barriers to effective communication and cultural integration.

Effective communication is vital in bridging civilian cultural gaps, but a lack of understanding can hinder this process. Veterans may struggle to articulate their experiences, military jargon, and unique perspectives in a way that civilians can comprehend. Similarly, civilians may have difficulty grasping the significance of military service, the challenges veterans face, and the sacrifices they have made. This lack of understanding can lead to miscommunication, frustration, and a failure to establish common ground. Overcoming communication challenges requires both veterans and civilians to actively seek understanding, listen attentively, and bridge the gap through shared experiences and empathy.

Stereotypes and preconceived notions about veterans and civilians further contribute to the lack of understanding between the two groups. Veterans may be perceived as rigid, authoritarian, or suffering from mental health issues, while civilians may be viewed as unaware of the sacrifices and commitment required in military service. These stereotypes can create barriers to forming genuine connections and hinder the process of closing cultural gaps. Overcoming these biases requires open-mindedness, willingness to challenge assumptions, and the recognition that individuals within both groups are diverse and multifaceted.

Developing empathy and practicing perspective-taking is essential in bridging the lack of understanding between veterans and civilians. Empathy allows individuals to understand and share the feelings and experiences of others, fostering a sense of connection and mutual respect. By actively seeking to understand each other's perspectives, veterans and civilians can bridge the gap and develop a deeper appreciation for their unique challenges, strengths, and contributions. This requires active listening, engaging in meaningful conversations, and recognizing the value of diverse experiences.

Education and awareness are key strategies in addressing the lack of understanding between veterans and civilians. By promoting cultural-competency training, workshops, and dialogue, both groups can gain insights into each other's cultures, values, and experiences. Veterans can share their stories, providing civilians with a better understanding of military life and the sacrifices made. Civilians, in turn, can educate veterans about civilian culture, societal norms, and expectations. Such initiatives facilitate empathy, dispel misconceptions, and create a more inclusive environment that encourages mutual understanding and respect.

Community engagement and integration play a vital role in closing civilian cultural gaps. Veterans can actively participate in community events, volunteer work, and social activities, creating opportunities for interaction and shared experiences with civilians. By engaging with the civilian community, veterans can demonstrate their skills, contribute to society, and break down barriers based on misunderstandings or stereotypes.

4. Employment and Career Transitions: Transitioning from military to civilian employment can be challenging due to differences in work cultures, expectations, and qualifications. Veterans may face difficulties in translating their military skills and experiences into civilian job requirements, and they may encounter bias or misconceptions about their abilities. It is essential to provide resources and support for veterans during the job-search process, such as career counseling, resume assistance, and networking opportunities. Employers can also benefit from education and training programs on the value of veterans' skills and experiences.

Transitioning from military service to a civilian career can be a challenging process for veterans, as they navigate cultural gaps that exist between the military and civilian sectors. Veterans bring valuable skills, experience, and discipline to the civilian workforce; but they may encounter barriers stemming from differences in organizational structure, communication styles, and workplace norms. The military operates under a well-defined hierarchical structure, with clear chains of command and strict adherence to protocols. In contrast, civilian organizations often have a more flattened organizational structure and emphasize collaborative decision-making. Veterans transitioning to civilian careers may initially find it challenging to adjust to a less rigid hierarchy. However, by proactively seeking opportunities to understand the civilian organizational structure, veterans can adapt and leverage their leadership skills and experience to contribute effectively in their new work environments.

Communication styles can differ significantly between military and civilian settings, which can create a gap during veterans' career transitions. The military relies on direct, concise, and task-oriented communication, while civilians often value more nuanced and collaborative communication approaches. Veterans may initially struggle to adapt to the more informal and open communication style prevalent in civilian workplaces. However, by actively observing and adapting to the communication norms of their new work environment, veterans can bridge the gap and effectively convey their ideas and expertise.

Civilian workplaces have unique norms and expectations that may differ from those experienced in the military. For instance, the military emphasizes punctuality, adherence to strict schedules, and a focus on the mission. In contrast, civilians may value flexibility, work-life balance, and individualism. Veterans may face challenges in adjusting their work habits and aligning with these new expectations. By being proactive in understanding and embracing civilian workplace norms, veterans can enhance their integration and effectiveness, while still leveraging their discipline and commitment to excellence.

One of the significant challenges veterans encounter during their career transition is conveying the transferability of their military skills and experience to civilian employers. Civilian employers may not always fully understand the value and applicability of the skills veterans have acquired, such as leadership, problem-solving, adaptability, and teamwork. Veterans can overcome this gap by clearly translating their military experience into relevant civilian terms during job interviews and on their resumes. Additionally, seeking out mentors, career counselors, and networking opportunities can provide valuable guidance and support in navigating this transition.

To bridge the cultural gaps in their career transition, veterans should prioritize ongoing professional development and training. This can include acquiring civilian certifications, attending workshops or courses to develop specific skills, and seeking opportunities for mentorship and networking. By investing in continuous learning, veterans can demonstrate their commitment to professional growth and adaptability in the civilian work environment. This proactive approach not only enhances their marketability but also helps bridge the cultural gaps by demonstrating their willingness to embrace new knowledge and practices.

Building cultural competence and sensitivity is essential for veterans navigating civilian career transitions. This involves seeking understanding, respecting diverse perspectives, and adapting to different cultural norms and practices. Veterans can engage in cultural competency training, diversity awareness programs, and cross-cultural workshops to develop the necessary skills and mindset. By demonstrating cultural sensitivity, veterans can bridge the gaps, foster inclusion, and contribute positively to their new workplaces.

5. Social Isolation and Loneliness: Veterans may experience social isolation and loneliness as they try to find their place in civilian communities. Military service fosters a strong sense of camaraderie and belonging, which may be challenging to replicate in civilian life. Veterans may feel disconnected from their peers or struggle to form new social relationships. Encouraging veterans to engage in community activities, join veteran organizations, or participate in support groups can help combat social isolation and foster a sense of belonging.

Social isolation is a significant challenge faced by veterans as they transition from military to civilian life. This isolation can be further exacerbated by the cultural gaps that exist between the military and civilian contexts. We aim to investigate the consequences of social isolation on veterans and its connection to fostering understanding and unity between military and civilian cultures. The transition from military service to civilian life often entails leaving behind a tightly-knit support network that veterans had within the military community. This loss can lead to a sense of social isolation, as veterans may find it challenging to recreate the same level of camaraderie and understanding in civilian contexts. The military culture emphasizes teamwork, shared experiences, and a strong sense of belonging, which may not be easily replicated in civilian life. As a result, veterans may struggle to find meaningful connections and a sense of purpose in their new civilian communities.

Social isolation among veterans can stem from the lack of shared experiences and understanding between the military and civilian cultures. Veterans have unique backgrounds, skills, and perspectives that may not align with those of civilians who have not served in the military. This disconnect can result in difficulties in relating to others, sharing personal stories, and finding common ground. As veterans navigate civilian cultural gaps, their experiences and sacrifices may be misunderstood or underappreciated, leading to feelings of isolation and alienation.

Communication barriers can contribute to social isolation among veterans. The military has its distinct jargon, acronyms, and communication styles that may not be readily understood by civilians. This can make it challenging for veterans to express themselves effectively and engage in meaningful conversations, resulting in feelings of frustration and isolation. Additionally, civilians may have limited exposure to military culture and may lack the knowledge or understanding to bridge the communication gap. Overcoming these barriers requires efforts from both veterans and civilians to actively listen, seek clarification, and cultivate mutual understanding.

Social isolation among veterans can be further exacerbated by the stigma surrounding mental health. Many veterans face mental health challenges such as PTSD, depression, and anxiety as a result of their military experiences. However, due to the perceived stigma, veterans may be hesitant to seek support or disclose their struggles. This can lead to feelings of isolation and exacerbate the challenges of bridging civilian cultural gaps. Creating a supportive and accepting environment that encourages open dialogue about mental health is crucial for addressing social isolation and promoting the well-being of veterans.

Efforts can be made to mitigate social isolation among veterans and bridge the gap between military and civilian cultures. Community-based programs and organizations that focus on veteran integration, peer support, and social activities can help foster connections and a sense of belonging. Mentoring programs that pair veterans with civilian mentors can also facilitate understanding, cultural exchange, and the development of meaningful relationships. Raising awareness about the experiences of veterans, promoting cultural sensitivity, and providing educational opportunities for civilians can foster empathy and reduce social isolation.

Collaboration and engagement between veterans and civilians are key to overcoming social isolation and bridging cultural gaps. Encouraging veterans to actively participate in community events, volunteer work, and social activities creates opportunities for interaction, shared experiences, and mutual understanding. Similarly, civilians can actively engage with veterans, showing support, appreciation, and a willingness to learn about their experiences. By working together, veterans and civilians can create an environment to bridge cultural gaps and mitigate social isolation for veterans.

The mental well-being of veterans plays a critical role in their ability to bridge the gaps that exist between military and civilian cultures. As veterans transition from the structured and disciplined environment of the military to the more fluid and diverse civilian world, they often encounter unique challenges that can impact their mental health. Our focus will be on examining the importance of veterans' mental well-being in bridging the gaps between civilian cultures, exploring the potential challenges they encounter, and identifying strategies to enhance their overall psychological health throughout the transition process.

Veterans' mental well-being is closely linked to their emotional resilience and adaptability. The military fosters a culture where veterans are trained to face adversity, manage stress, and cope with challenging situations. However, the transition to civilian life can present emotional challenges that veterans may not have previously encountered. Maintaining good mental health allows veterans to navigate the emotional complexities of civilian culture gaps, adapt to new environments, and effectively interact with civilians. Emotional resilience helps veterans approach cultural differences with an open mind, curiosity, and empathy, fostering understanding and connection.

Positive mental well-being is crucial for veterans to engage in effective communication and build relationships in civilian settings. The ability to express oneself clearly, manage emotions, and actively listen is essential in bridging cultural gaps. Veterans who prioritize their mental well-being can develop stronger emotional intelligence, which enhances their communication skills and enables them to navigate sensitive topics, address misconceptions, and build meaningful connections with civilians. By fostering healthy relationships, veterans can create an environment where cultural understanding can flourish.

Feelings of loss, isolation, and identity transformation can arise as veterans adjust to the absence of the military structure and camaraderie they were accustomed to. These challenges can be further intensified by the cultural gaps they encounter. Prioritizing mental well-being allows veterans to develop effective coping mechanisms, seek support when needed, and build resilience in the face of these challenges. By addressing their mental health needs, veterans are better equipped to overcome obstacles and engage meaningfully in the process of closing civilian culture gaps.

Promoting veterans' mental well-being involves addressing the stigma often associated with seeking support for psychological concerns. Encouraging veterans to prioritize their mental health and providing accessible resources helps create a culture of support and understanding. Veterans who prioritize their mental well-being can seek counseling, therapy, or support groups, which provide a safe space to navigate the complexities of the transition and address any emotional challenges they may face. By breaking down the barriers to seeking support, veterans can strengthen their psychological resilience, which positively impacts their ability to bridge civilian cultural gaps.

Veterans' mental well-being directly influences their capacity for empathy and cultural understanding. When veterans prioritize their mental health, they can better manage stress, regulate their emotions, and maintain a positive outlook. This emotional stability allows veterans to approach cultural gaps with empathy, patience, and an open mind. By fostering their well-being, veterans can develop a deeper understanding of the challenges faced by both themselves and civilians during the transition. This understanding promotes mutual respect, enhances cultural integration, and closes the gaps that exist between military and civilian cultures.

The mental well-being of veterans is a crucial factor in their ability to bridge civilian cultural gaps. By prioritizing mental health, veterans can develop emotional resilience, effective communication skills, and coping mechanisms that facilitate cultural understanding and connection. A focus on mental well-being not only supports veterans' transition to civilian life but also promotes healthy psychological practices that veterans can use throughout their lives.

Navigating civilian culture gaps is a significant struggle for veterans transitioning from military service. The challenges they face with communication, differences in values, misconceptions, employment transitions, social isolation, and mental health require understanding and support from both veterans and civilians. By fostering awareness, promoting dialogue, providing resources, and creating inclusive environments, we can bridge the gaps and create a more supportive and welcoming society for our veterans. Ultimately, recognizing and addressing these struggles will help veterans successfully reintegrate into civilian life and thrive in their postmilitary journeys.

Civilian cultural gaps have a significant impact on veterans' ability to effectively translate their military skills into the civilian context. Military service provides veterans with a unique set of skills and experiences that are highly valuable in various industries. However, the differences in organizational structures, communication styles, and expectations between the military and civilian sectors can present challenges when veterans try to articulate and apply their military skills in civilian settings.

Effective communication is crucial for veterans to translate their military skills, but civilian cultural gaps can impact veterans' ability to communicate in civilian settings. The military emphasizes direct and concise communication, while civilians often value more nuanced and diplomatic communication styles. Veterans may struggle to adapt their communication approach to align with civilian expectations, potentially leading to misunderstandings or misinterpretations of their abilities. Overcoming this gap requires veterans to actively listen, observe communication norms in civilian environments, and adapt their communication styles accordingly.

Cultural gaps can also influence the recognition and understanding of veterans' transferable skills by civilian employers. While veterans possess a wide range of skills such as leadership, problem-solving, teamwork, and resilience, these skills may not be immediately recognizable or understood by civilians who lack familiarity with military experiences. Veterans must highlight and effectively demonstrate how their military skills can be applied to specific roles or industries. This can be achieved through targeted resumes, cover letters, and interviews, where veterans can translate their military experiences into civilian terms and showcase the value they bring to the civilian workplace.

Cultural competency and education play a crucial role in bridging civilian cultural gaps and facilitating the translation of military skills. Employers and organizations can promote cultural competence by providing training, resources, and awareness programs that educate civilian personnel about the unique experiences, skills, and challenges veterans bring. Similarly, veterans can seek out opportunities for professional development, certifications, and additional education that enhance their understanding of civilian culture and industry-specific knowledge. By actively pursuing cultural competency and education, both veterans and civilians can foster a more inclusive and supportive environment that facilitates the translation of military skills.

Chapter 3

Translating Military Skills

Translating military skills for veterans is of utmost importance in facilitating their successful transition from military service to civilian life. Veterans possess a wealth of unique and valuable skills acquired during their time in the military, ranging from leadership and teamwork to problem-solving and adaptability. However, these skills may not always be easily understood or recognized by civilian employers and industries. The process of translating military skills involves effectively communicating the relevance and transferability of these skills in a civilian context. In this chapter, we highlight the importance of translating military skills for veterans and its impact on their postmilitary career success.

1. Unlocking Employment Opportunities: Translating military skills is vital for unlocking a wide range of employment opportunities for veterans. Many veterans possess highly specialized skills and experiences that are highly sought after in various industries, such as logistics, engineering, cybersecurity, and project management. However, the language and terminology used in the military may not directly align with civilian job descriptions or requirements. By effectively translating military skills into civilian language, veterans can showcase their qualifications, capabilities, and experiences to potential employers. This opens doors to diverse career paths and increases their chances of finding meaningful employment that aligns with their abilities and interests.

Transitioning from military service to the civilian workforce can present significant challenges for veterans. However, unlocking employment opportunities for veterans is a critical aspect of supporting their successful integration into civilian life. Veterans possess a wide range of valuable skills, experiences, and qualities acquired during their military service that can greatly contribute to various industries and organizations. One of the key steps in unlocking employment opportunities for veterans is recognizing the value of their military skills. The military provides extensive training in technical areas, leadership, teamwork, problem-solving, adaptability, and discipline among others. These skills are highly transferable to civilian occupations and are in high demand across various industries. By acknowledging and appreciating the skills that veterans bring, employers can tap into a pool of qualified and experienced candidates who possess unique perspectives and a strong work ethic. This recognition sets the stage for unlocking employment opportunities by creating a positive perception of veterans' capabilities and their potential contributions to organizations.

Translating military skills into civilian language is a crucial step in unlocking employment opportunities for veterans. While veterans possess a wealth of skills and experiences, the terminology used in the military may not directly align with civilian job descriptions or requirements. It is essential to bridge this gap by effectively translating military skills to highlight their relevance and transferability in a civilian context. By articulating how their military experiences have equipped them with skills that align with specific job requirements, veterans can increase their chances of being considered for employment. This translation process involves communicating the value and applicability of military skills in a way that resonates with potential employers, thereby unlocking a broader range of job opportunities.

Building networks and partnerships is another important aspect of unlocking employment opportunities for veterans. Collaborating with organizations, businesses, and community groups that have a vested interest in supporting veterans' employment can provide valuable connections and resources. These partnerships can offer job placement assistance, mentoring programs, networking events, and educational opportunities that specifically cater to veterans' needs. By leveraging these networks, veterans gain access to a wider range of employment opportunities and are more likely to connect with employers who value their military experience. Additionally, networking enables veterans to tap into the hidden job market where many positions are filled through personal connections rather than traditional job postings.

Tailoring resumes and cover letters to highlight relevant military skills and experiences is crucial in unlocking employment opportunities for veterans. Resumes should emphasize the specific accomplishments, qualifications, and responsibilities that directly align with the desired civilian roles. By customizing their application materials to emphasize the transferable skills gained in the military, veterans can capture the attention of employers and stand out from other applicants. Additionally, including a well-written cover letter that communicates how their military experience relates to the specific job requirements can further enhance their chances of securing employment opportunities.

To unlock employment opportunities for veterans, it is essential to provide them with training and educational opportunities that bridge any skill gaps and facilitate their transition to civilian careers. Some veterans may require additional education, certification, or licensing to meet the specific requirements of certain industries or professions. By offering targeted training programs, educational grants, or support for professional certifications, veterans can acquire the necessary qualifications to access a broader range of employment opportunities. Providing ongoing professional development opportunities also ensures that veterans remain competitive in the evolving job market.

Promoting employer engagement and incentives for veterans is crucial in unlocking employment opportunities and supporting their successful transition to civilian careers. Employers play a pivotal role in recognizing and appreciating the skills, experiences, and qualities that veterans bring to the workforce. By actively engaging with employers, organizations, and industry leaders, it is possible to create awareness and promote the benefits of hiring veterans. This can be achieved through outreach programs, networking events, and employer-training sessions that emphasize the value of military skills and the positive impact veterans can have on organizational success. Additionally, providing incentives such as tax credits, grants, or special recruitment initiatives can further encourage employers to prioritize the hiring of veterans. By fostering employer engagement and incentives, we can ensure that veterans are given the opportunities they deserve to contribute their unique talents and experiences to the civilian workforce.

2. Demonstrating Transferable Skills: Translating military skills allows veterans to demonstrate the transferability of their capabilities to civilian roles. While the specific tasks performed in the military may differ from those in civilian jobs, the underlying skills acquired through military training are often highly transferable. For example, the ability to work under pressure, make critical decisions, communicate effectively, and lead teams are all valuable skills in both military and civilian contexts. By effectively translating and highlighting these transferable skills, veterans can bridge the gap between their military experience and the requirements of civilian employers. This helps employers recognize the value veterans bring to their organizations, leading to increased opportunities for employment and career advancement.

One of the key challenges veterans face during their transition from military service to the civilian workforce is effectively demonstrating their transferable skills. Veterans possess a wide range of valuable skills acquired through their military training and experience, but these skills may not always be immediately recognizable or understood by civilian employers. The ability to articulate and showcase transferable skills is essential in bridging the gap between military and civilian employment.

The first step in demonstrating transferable skills is understanding what they are and how they can be applied in a civilian context. Transferable skills are qualities and abilities that can be utilized across different roles, industries, and settings. They are not specific to a particular job or profession but rather represent the broader capabilities developed through military training and service. Examples of transferable skills include leadership, teamwork, problem-solving, adaptability, communication, organization, and time management. By recognizing and acknowledging their transferable skills, veterans can identify the valuable attributes they possess and how these can be applied in civilian employment.

Once veterans understand their transferable skills, the next step is to identify relevant experiences that showcase these skills in action. Military service provides numerous opportunities for veterans to develop and demonstrate transferable skills. Whether it's leading a team during a mission, managing resources, coordinating logistics, or adapting to rapidly changing circumstances, veterans can draw on specific experiences to highlight their skills and competencies. By identifying and reflecting on these experiences, veterans can effectively articulate how their military background has equipped them with the necessary skills for success in civilian employment.

To demonstrate transferable skills effectively, veterans must tailor their resumes and cover letters to emphasize the specific skills and experiences relevant to the desired civilian roles. When crafting their resumes, veterans should focus on describing their accomplishments, responsibilities, and contributions in a way that highlights their transferable skills. This involves using civilian-friendly language and terminology that potential employers can easily understand. Similarly, the cover letter should provide a concise summary of how the transferable skills acquired in the military directly align with the requirements of the job. By tailoring their application materials, veterans can effectively demonstrate the value they bring to potential employers and increase their chances of securing employment opportunities.

When discussing transferable skills in interviews or networking events, veterans should provide concrete examples to support their claims. Instead of simply stating that they possess strong leadership skills, veterans should share specific instances where they led a team, made critical decisions under pressure, or achieved positive outcomes through effective leadership. By sharing real-life examples, veterans can showcase their transferable skills in action and provide tangible evidence of their abilities. These examples not only validate the skills but also help potential employers visualize how they can be applied in a civilian work environment.

In some cases, veterans may need to seek additional professional development or training to further enhance their transferable skills.

Depending on the desired civilian career path, acquiring specific certifications, licenses, or qualifications may be necessary. Veterans should proactively identify areas where they may need to upskill or acquire new knowledge to align their transferable skills with the requirements of their chosen field. Participating in professional development programs, attending workshops, or pursuing further education can significantly strengthen their transferable skills and increase their employability in the civilian workforce.

Leveraging networking and mentoring is instrumental for veterans as they navigate their transition from military service to the civilian workforce. Networking allows veterans to connect with professionals in their desired industries, gain industry insights, and create valuable connections that can lead to employment opportunities. By attending networking events, joining professional organizations, and utilizing online platforms, veterans can expand their professional network and increase their visibility to potential employers. Mentoring relationships, on the other hand, provide veterans with guidance and support from experienced professionals who can share their knowledge, offer career advice, and help veterans navigate the nuances of the civilian workplace. These mentors can provide valuable insights into the industry, assist with resume and interview preparation, and serve as advocates for veterans in their career pursuits. Leveraging networking and mentoring opportunities empowers veterans to build meaningful connections, access valuable resources, and enhance their chances of securing employment and achieving long-term career success.

3. Enhancing Career Development: Translating military skills plays a crucial role in enhancing veterans' career development and growth. The process of transitioning from military service to civilian employment is challenging, particularly when veterans struggle to articulate the relevance and applicability of their military skills. By effectively translating their skills, veterans can align their experiences with specific career paths, educational opportunities, and professional-development programs. This not only enables veterans to pursue their career goals but also promotes their ongoing growth and success

in civilian occupations. By recognizing the transferability of their skills, veterans can make informed decisions about career transitions, identify areas for further skill development, and seek out opportunities that leverage their unique strengths.

Career development is a critical aspect of supporting veterans in their transition from military service to civilian careers. By providing opportunities and resources for career growth and advancement, veterans can effectively navigate the challenges of transitioning into the civilian workforce and achieve long-term professional success. The first step in enhancing career development for veterans is recognizing and valuing their transferable skills and experiences gained during their military service. Veterans possess a wide range of valuable skills such as leadership, teamwork, problem-solving, adaptability, and discipline, which are highly transferable to various industries and professions. By acknowledging and leveraging these skills, employers and organizations can provide opportunities for veterans to apply and develop their capabilities in civilian roles. Recognizing the unique strengths that veterans bring to the table sets the foundation for career-development initiatives tailored to their specific needs and aspirations.

To enhance career development for veterans, it is crucial to provide targeted training and educational opportunities that align with their career goals and fill any skill gaps they may have. Some veterans may require additional education, certification, or licensing to meet the specific requirements of their desired industries or professions. Offering vocational training programs, educational grants, or support for professional certifications can empower veterans to acquire the necessary qualifications and competencies for career advancement. Continuous learning and upskilling enable veterans to remain competitive in the ever-evolving job market and adapt to emerging trends and technologies.

Mentoring and coaching programs play a vital role in enhancing career development for veterans. Pairing veterans with experienced professionals in their desired fields provides them with guidance, support, and insights into the civilian workplace. Mentors can offer career advice, share their experiences, and provide valuable networking opportunities. These relationships can help veterans set clear career goals, identify growth opportunities, and develop the necessary skills and knowledge to progress in their chosen fields. Coaching programs can also provide structured support, assisting veterans in setting objectives, overcoming challenges, and navigating career transitions.

Creating a culture of internal mobility and advancement is essential for enhancing career development for veterans. Organizations should prioritize the identification and development of high-potential veterans, providing them with opportunities to take on new challenges, assume leadership roles, and expand their skill sets. Promoting from within not only recognizes the talent and potential of veterans, but also fosters a sense of loyalty and commitment to the organization. Implementing clear career progression paths, mentorship programs, and performance management systems that align with veterans' career goals can provide a structured framework for their growth and advancement.

Transition support services play a crucial role in enhancing career development for veterans by addressing the unique challenges they may face during the transition process. These services may include job-placement assistance, resume and interview preparation, networking opportunities, and career counseling. By offering comprehensive support tailored to veterans' needs, organizations can help them overcome barriers, gain confidence, and navigate the civilian job market more effectively. Providing transitional support services also demonstrates a commitment to veterans' career development and sends a positive message about the value placed on their contributions and growth within the organization.

Encouraging veterans to actively participate in professional networking and engagement opportunities can significantly enhance their career development. Networking events, industry conferences, and veteran-specific organizations provide platforms for veterans to connect with professionals in their fields of interest, gain industry insights, and explore new career opportunities. By building and nurturing professional relationships, veterans expand their network, access job openings, and gain valuable advice and guidance from experienced professionals.

4. Bridging the Cultural Gap: Translating military skills helps bridge the cultural gap between the military and civilian

sectors. Translating military skills involves understanding and adapting to the cultural nuances of the civilian workforce. By effectively communicating their skills and experiences in terms that resonate with civilian employers, veterans can bridge the cultural gap, demonstrate their readiness to contribute, and integrate more seamlessly into civilian workplaces. This leads to a smoother transition, improved workplace relationships, and enhanced job satisfaction for veterans.

The military and civilian cultures can differ significantly in terms of values, language, work environment, and expectations. This cultural divide can pose unique challenges for veterans as they navigate the civilian workforce and strive to integrate into a new social and professional environment. To bridge the culture gap, it is essential to enhance cultural awareness among both veterans and the civilian population. Veterans should be provided with information and resources to understand the civilian workplace culture, its norms, and expectations. Likewise, employers, colleagues, and communities should be educated about the military culture, its values, and the experiences of veterans. Cultural awareness training can promote mutual understanding, break down stereotypes, and foster a more inclusive environment. By developing a shared understanding of each other's backgrounds and experiences, veterans and civilians can build stronger relationships and work collaboratively.

Transition support programs play a vital role in bridging the culture gap for veterans. These programs offer assistance and guidance to veterans as they navigate the civilian workforce, helping them understand the cultural nuances and expectations in the civilian workplace. Transition support may include workshops, seminars, or one-on-one counseling sessions, providing veterans with information on workplace etiquette, communication styles, dress codes, and other cultural aspects. By equipping veterans with the necessary knowledge and skills, these programs facilitate a smoother transition and enable veterans to adapt to the new culture more effectively.

Encouraging cross-cultural collaboration is crucial for bridging the cultural gap. Employers and organizations should create opportunities for veterans and civilians to work together on projects, fostering mutual understanding and respect. Collaborative team-building activities, diversity training, and cross-functional projects can bring together individuals from different backgrounds, facilitating meaningful interactions and breaking down cultural barriers. By promoting a culture of inclusivity and collaboration, organizations can create an environment where veterans feel valued, supported, and able to contribute their unique perspectives and skills.

Employee resource groups (ERGs) can play a significant role in bridging the culture gap for veterans. These groups provide a platform for veterans to connect, share experiences, and support each other in the civilian workplace. ERGs also promote awareness and understanding of military culture among the civilian workforce by organizing educational events, cultural celebrations, and community-service initiatives. Through ERGs, veterans can find a sense of community, seek guidance, and develop strategies to navigate the cultural transition. Supporting and encouraging the establishment of ERGs demonstrates an organization's commitment to fostering a supportive and inclusive environment for veterans.

Promoting cultural competence training for veterans is crucial in ensuring their successful integration into the civilian workforce. Cultural competence training provides employers, colleagues, and communities with a deeper understanding of the unique experiences, values, and challenges faced by veterans. By participating in this training, individuals gain insight into military culture, its language, customs, and the impact of military service on veterans' lives. This training helps foster empathy, sensitivity, and awareness, enabling a more inclusive and supportive environment for veterans. By promoting cultural competence, organizations and communities can bridge the culture gap; promote effective communication; and create an atmosphere where veterans feel understood, valued, and respected in their transition to civilian life.

5. Advocacy and Appreciation: Translating military skills promotes advocacy and appreciation for the unique talents and experiences veterans bring to the civilian workforce. Veterans

possess a breadth of skills and qualities developed through their military service, such as discipline, resilience, teamwork, and adaptability. By translating these skills and highlighting their relevance, employers and the broader community gain a deeper understanding of the value veterans offer. This fosters a culture of appreciation where veterans' contributions are recognized, respected, and valued.

Advocacy and appreciation for veterans play a vital role in recognizing their service, supporting their transition to civilian life, and ensuring their well-being. Veterans have made significant sacrifices to protect and serve their country, and society must acknowledge and values their contributions. Our research will center around the significance of advocating for and appreciating veterans while also exploring how their contributions can enhance the overall well-being and success of the veteran community.

One of the fundamental aspects of advocating for veterans is recognizing and honoring their service. Veterans have demonstrated courage, dedication, and a commitment to protecting their nation. Publicly acknowledging their sacrifice and expressing gratitude can have a profound impact on their sense of worth and belonging. Veterans Day ceremonies, parades, and other commemorative events provide opportunities for communities to come together, show support, and express appreciation for the sacrifices veterans have made. Such recognition not only validates veterans' experiences but also reinforces the understanding that their service is valued and respected by society.

Advocacy for veterans also involves providing support during their transition from military service to civilian life. This support can include assistance with job placement, educational opportunities, housing, health care, and mental health services. Recognizing the unique challenges veterans face, such as translating military skills, navigating the civilian job market, and adjusting to a different lifestyle, is crucial. Advocacy efforts can focus on establishing programs and initiatives that address these specific needs, ensuring that veterans have the resources and support to successfully reintegrate into civilian society. By offering tailored transition support, society demonstrates its commitment to veterans' well-being and recognizes the importance of their successful reintegration.

The prioritization of mental health and well-being should take precedence in advocating for veterans. The experiences of military service can leave a lasting impact on veterans, and many may face mental health challenges such as PTSD, depression, and anxiety. It is essential to destigmatize mental health issues and provide accessible and comprehensive mental health services for veterans. Advocacy efforts can focus on raising awareness, reducing barriers to mental health care, and ensuring that veterans have the necessary support systems in place. This can involve collaborations between government agencies, nonprofit organizations, and health-care providers to establish dedicated programs that address the mental health needs of veterans. By advocating for mental health support, society demonstrates its commitment to the overall well-being and quality of life of veterans.

In addition to mental health and well-being, advocacy for veterans should center on the creation of employment and economic opportunities. Many veterans possess valuable skills and experiences that can contribute to the civilian workforce, but they may face challenges in translating their military skills and finding suitable employment. Advocacy efforts can involve working with employers to establish veteran-friendly hiring practices, encouraging the recognition of military training and experience, and providing resources for veterans to develop their professional skills and access job opportunities. Additionally, advocating for entrepreneurship and small business support can empower veterans to start their businesses and contribute to the local economy. By advocating for employment and economic opportunities, society helps veterans build successful civilian careers, support their families, and contribute to the growth and prosperity of their communities.

Advocacy for veterans should also prioritize education and training opportunities. Many veterans may seek to pursue higher education or acquire new skills to enhance their career prospects in the civilian workforce. Advocacy efforts can involve promoting educational grants, scholarships, and programs specifically tailored to veterans' needs. Collaborations between educational institutions and veterans' organizations can help streamline the transition process, provide academic support, and create a supportive learning environment for veterans. By advocating for education and training opportunities, society demonstrates its commitment to lifelong learning, professional development, and empowering veterans to achieve their full potential in their civilian goals.

Translating military skills into civilian terms is a critical step for veterans during their transition to the civilian workforce. While it is essential for showcasing their qualifications and abilities, the process of translating military skills can sometimes have unintended consequences, including limitations on networking capabilities.

One of the primary effects of translating military skills is the creation of language and terminology barriers that can hinder effective networking. The military has its distinct language and acronyms that may not readily align with the language used in civilian industries. When veterans attempt to communicate their skills and experiences in civilian terms, there can be a loss of specificity and nuance, making it difficult for civilians to fully grasp the depth and breadth of their capabilities. This language barrier can hinder meaningful conversations and connections during networking events as civilians may struggle to fully understand the unique skills and experiences that veterans bring to the table.

Another effect of translating military skills is the potential for misunderstandings and the perpetuation of stereotypes. Veterans may find it challenging to articulate their military experiences and how they relate to civilian contexts, leading to misunderstandings or misperceptions from civilians. These misunderstandings can contribute to the perpetuation of stereotypes, such as assumptions that veterans are solely suited for certain roles or lack relevant skills for civilian careers. As a result, veterans may face limited networking opportunities as others may not fully recognize or appreciate the range of transferable skills they possess. Overcoming these stereotypes and misconceptions requires proactive efforts to educate civilians about the diverse skill sets and experiences veterans bring to the civilian workforce.

Chapter 4

Limited Civilian Networks

Veterans may encounter distinctive challenges during the transition from military to civilian life, which can include limitations in their ability to establish civilian networks. While veterans possess valuable skills and experiences, certain factors can hinder their ability to effectively network and establish connections in the civilian world. Within this chapter, we will examine several prevalent ways through which veterans might unintentionally curtail their ability to develop civilian networks while going through the transition process.

1. Failing to Translate Military Skills: One of the most common ways veterans limit their civilian networking capabilities is by failing to effectively translate their military skills and experiences into language that resonates with civilian professionals. The military has its terminology, acronyms, and jargon, which may not be readily understood by those outside the military community. When veterans attempt to communicate their skills and experiences using military language, it can create a barrier to effective networking. Employers and professionals from different industries may struggle to fully grasp the relevance and applicability of the veterans' skill set, leading to missed networking opportunities. To overcome this limitation, veterans must invest time and effort in translating their military skills into civilian terms, highlighting transferable abilities that align with the needs of the industry or profession they are targeting.

One of the primary hurdles veterans face is the difficulty in translating their military skills into the language and terminology familiar to civilian employers. Military jargon and acronyms, which are deeply ingrained in the military culture, can create a communication barrier when trying to convey their qualifications to civilian hiring managers. This lack of alignment in language and terminology can lead to a misunderstanding of the valuable skills and experiences veterans bring to the table, limiting their chances of securing suitable employment opportunities.

Another challenge lies in the perception that military skills may not be directly transferable to civilian roles. Employers may struggle to recognize the applicability and relevance of veterans' experiences, assuming that they are only suited for military-specific jobs. This narrow perspective can result in missed opportunities for veterans to showcase their transferable skills, adaptability, leadership abilities, and problem-solving aptitude, which are all highly valuable in various professional settings.

Veterans often find it challenging to navigate the nuances of the civilian work culture, which can be distinct from the structured and hierarchical military environment. This disparity can manifest in various ways, including difficulty in adapting to different communication styles, workplace dynamics, and expectations. As a result, veterans may inadvertently face obstacles in networking, building professional relationships, and assimilating into the civilian work environment, hindering their career progression and integration into new industries.

In certain occupations, the lack of civilian credentials or certifications can pose a significant barrier for veterans. Despite possessing extensive hands-on experience and practical skills, veterans may encounter challenges in meeting the specific licensing or certification requirements demanded by civilian employers. This disparity in credentialing can limit their access to certain career paths, impeding their ability to fully leverage their military expertise and limiting their opportunities for advancement.

Successfully translating military skills into the civilian job market is a critical aspect of veterans' successful transition into civilian life. The challenges faced by veterans in this process, including language barriers, nontransferable perceptions, limited exposure to civilian work culture, and insufficient civilian credentialing, hinder their ability to effectively showcase their talents and experiences. Addressing these challenges requires a collaborative effort involving employers, government agencies, educational institutions, and veterans' support organizations to bridge the gap and create more opportunities for veterans to thrive in the civilian job market. By recognizing and valuing the unique skills veterans possess and providing the necessary resources and support, we can empower veterans to overcome these obstacles and unlock their full potential in civilian careers.

2. Overemphasizing Military Identity: Another way veterans may inadvertently limit their networking capabilities is by overemphasizing their military identity to the point of overshadowing other aspects of their professional persona. While military service is a significant part of a veteran's life and should be acknowledged, solely focusing on military experiences during networking interactions can hinder connections with civilians who may not fully understand or relate to those experiences. Veterans should strive to strike a balance between highlighting their military background and showcasing their broader skill set and potential contributions in a civilian context. This approach allows for a more well-rounded networking experience and facilitates connections beyond the military sphere.

When veterans transition out of the military and enter the civilian world, they must strike a balance in how they present their military identity. While military service is a significant part of a veteran's life and should be acknowledged, overemphasizing their military identity can sometimes have unintended consequences and limit their opportunities for personal and professional growth. Overemphasizing their military identity can hinder veterans' ability to expand their professional identity beyond the confines of their military experience. Veterans bring a wealth of skills, knowledge, and abilities acquired through their military service; but they need to recognize and articulate the broader range of talents they possess. By focusing solely on their military background, veterans may unintentionally limit their potential for exploring new career paths, industries, or roles that align with their interests and aspirations. Embracing a more holistic professional identity allows veterans to tap into their full potential, pursue diverse opportunities, and adapt to the evolving demands of the civilian job market.

Expanding one's professional identity beyond the military is of utmost importance for military members as they transition into civilian life. While military service provides individuals with a unique skill set and valuable experiences, they need to recognize the significance of broadening their professional identity beyond the confines of their military careers.

By broadening their professional identity beyond the military, military members can fully embrace and highlight their transferable skills. The military equips individuals with a diverse range of skills, including leadership, teamwork, problem-solving, adaptability, and resilience. These skills are highly valuable in the civilian workforce, but military members need to recognize how these skills can be applied across different industries and professions. By expanding their professional identity, military members can effectively communicate their transferable skills to employers and demonstrate their versatility and readiness to contribute in various contexts.

Diversifying their professional identity beyond the military unlocks opportunities for veterans to explore new career paths that resonate with their interests and aspirations, thereby broadening their horizons and growth potential. Military members possess a wide range of talents and interests beyond their specific military occupation. By exploring different industries, job roles, and sectors, they can discover new passions and expand their skill set. This exploration allows military members to leverage their unique experiences and perspectives in a civilian context, fostering personal growth and enabling them to contribute meaningfully in areas they are passionate about.

By expanding their professional identity beyond the military, military members have the chance to cultivate a diverse professional network, enabling them to connect with individuals from various industries and backgrounds. While military networks are valuable and provide a sense of camaraderie and support, military members need to connect with professionals from various industries and backgrounds. Building a diverse network enables military members to access a wider range of resources, knowledge, and career opportunities. It also facilitates the exchange of ideas, collaboration, and mentorship as they engage with professionals who offer different perspectives and experiences.

Through the expansion of their professional identity beyond the military, military members bolster their adaptability and resilience, acquiring valuable skills and experiences that enable them to navigate diverse and dynamic environments with greater flexibility and strength. The military operates within a unique structure and culture with clear expectations and defined roles. Transitioning to the civilian workforce requires individuals to adapt to different organizational structures, work environments, and expectations. By expanding their professional identity, military members develop the skills and mindset necessary to navigate change, embrace new challenges, and thrive in diverse settings. This adaptability and resilience are valuable assets that enhance their long-term career prospects and enable them to succeed in an ever-evolving professional landscape.

By broadening their professional identity beyond the military, military members foster an environment that supports their personal growth and overall well-being, nurturing their individual development and contributing to a fulfilling and balanced life. Engaging in activities and pursuing career paths beyond the military can provide a sense of fulfillment and purpose. It allows military members to explore their interests, passions, and values, promoting a sense of self-discovery and personal satisfaction. Additionally, expanding their professional identity can help prevent feelings of isolation or loss of identity often experienced during the transition to civilian life. It allows military members to build new connections, develop a sense of belonging, and cultivate a well-rounded identity that extends beyond their military service.

Expanding their professional identity outside the military is crucial for military members as they transition into civilian life. By embracing their transferable skills, exploring new career paths, building a diverse professional network, enhancing adaptability and resilience, and fostering personal growth, military members can unlock a world of opportunities and thrive in their postmilitary careers. It is a transformative process that empowers military members to harness their full potential.

Overemphasizing their military identity can also hinder veterans' ability to adapt and be flexible in a civilian environment. The military operates within a highly structured and hierarchical system with its own set of protocols, procedures, and expectations. While these qualities are valuable in certain contexts, the civilian world often requires individuals to be adaptable, open-minded, and willing to navigate ambiguity. By broadening their identity beyond the military, veterans can demonstrate their ability to thrive in different environments, handle diverse challenges, and bring fresh perspectives to their professional endeavors. This flexibility enhances their marketability and enables them to seize new opportunities as they arise.

When veterans transition out of the military and enter the civilian world, one of the key skills they need to cultivate is adaptability and flexibility. To successfully navigate this transition, veterans must enhance their adaptability and flexibility, enabling them to embrace new challenges, thrive in diverse settings, and seize opportunities for personal and professional growth. Within the scope of this chapter, we will investigate the strategies and approaches available to veterans, which can be employed to enhance their adaptability and flexibility during the transition process.

Adopting a growth mindset is crucial for enhancing adaptability and flexibility. A growth mindset is a belief that abilities and intelligence can be developed through effort and learning. By cultivating a mindset that embraces continuous learning and improvement, veterans can adapt to new situations, acquire new skills, and overcome challenges with resilience. Embracing a growth mindset allows veterans to see the transition as an opportunity for growth and development rather than an insurmountable obstacle. It encourages them to view setbacks as learning experiences and seek out new opportunities for personal and professional advancement.

To enhance adaptability and flexibility, veterans should actively seek out new experiences that challenge their comfort zones. This can involve volunteering, participating in community activities, joining professional organizations, or pursuing education and training in new fields. Engaging in diverse experiences exposes veterans to different perspectives, cultures, and work environments, fostering adaptability and flexibility. It helps them develop the ability to quickly adjust to new situations, communicate effectively with individuals from various backgrounds, and adapt their skills to different contexts.

Developing a supportive network is crucial for veterans during the transition process. By connecting with fellow veterans, mentors, and professionals from various industries, veterans can gain valuable insights and support. These connections provide opportunities for learning, collaboration, and mentorship, helping veterans navigate the civilian world and adapt to new environments. Building a diverse network allows veterans to tap into different perspectives, experiences, and resources, further enhancing their adaptability and flexibility as they receive guidance and support from individuals with varied backgrounds.

Emotional intelligence plays a vital role in enhancing adaptability and flexibility. By cultivating self-awareness, empathy, and effective interpersonal skills, veterans can navigate complex social dynamics, build relationships, and collaborate effectively with diverse individuals. Understanding their own emotions and those of others allows veterans to adapt their communication styles, resolve conflicts, and build trust, ultimately enhancing their ability to navigate unfamiliar environments and succeed in their professional endeavors.

Enhancing adaptability and flexibility is essential for veterans during their transition out of the military. By adopting a growth mindset, seeking new experiences, developing transferable skills, building a supportive network, and cultivating emotional intelligence, veterans can successfully adapt to the civilian world and seize opportunities for personal and professional growth. These skills not only enhance their adaptability and flexibility but also equip them with a strong foundation for long-term success in their postmilitary careers.

3. Building Relatable Connections: Overemphasizing their military identity may also limit veterans' ability to build relatable connections with civilians who may not fully understand or relate to military experiences. Veterans have a wealth of valuable skills and experiences that can benefit various industries and sectors,

> but they must establish connections and build relationships with individuals outside the military community. By highlighting their broader skill set, veterans can establish common ground with civilians, foster understanding, and bridge the gap between military and civilian cultures. This opens doors for collaboration, mentorship, and support from professionals who may offer valuable insights, opportunities, and guidance in the civilian realm.

When veterans transition out of the military and into civilian life, they need to build relatable connections to their service. These connections can provide a sense of belonging, support, and understanding as veterans navigate the challenges and opportunities of their postmilitary lives. Building relatable connections allows veterans to maintain a connection to their military experiences while also integrating into civilian communities. Building relatable connections to their service provides veterans with a sense of belonging. The military community often fosters a strong bond among service members, built on shared experiences, values, and challenges. Transitioning out of the military can sometimes lead to a sense of isolation or a loss of identity for veterans. By building relatable connections, veterans can find a supportive network of individuals who have also served and understand the unique aspects of military life. These connections create a sense of camaraderie and belonging, providing veterans with a support system that can offer encouragement, empathy, and a shared sense of purpose.

By establishing relatable connections to their service, veterans find a supportive environment where their experiences, perspectives, and emotions are comprehended and affirmed. The civilian world may not fully comprehend the complexities and sacrifices of military service. However, connecting with other veterans allows for a mutual understanding of the challenges faced, the values upheld, and the impact of military experiences on individuals' lives. Veterans can share their stories, discuss their unique perspectives, and find validation in the shared understanding of fellow veterans. This validation reinforces veterans' self-worth and strengthens their sense of identity beyond their military service.

Establishing relatable connections provides veterans with valuable opportunities for peer support and mentorship. Veterans who have successfully transitioned out of the military can offer guidance, advice, and mentorship to those who are embarking on their transition journey. Peers who have shared similar experiences can provide valuable insights, practical tips, and emotional support to help veterans navigate the challenges of civilian life. Mentorship relationships allow veterans to learn from those who have walked the path before them, benefiting from their wisdom, experiences, and networks.

Relatable connections offer veterans opportunities for continued learning and personal growth. By connecting with other veterans, veterans' organizations, and support groups, veterans can access resources, educational programs, and professional-development opportunities tailored to their specific needs and interests. These connections foster a culture of lifelong learning, allowing veterans to acquire new skills, expand their knowledge, and adapt to the changing demands of the civilian world. By staying connected to their service through relatable connections, veterans can continue to grow and thrive in their postmilitary lives.

The act of building relatable connections acts as a crucial bridge connecting the military and civilian realms. Veterans can bring their unique skills, experiences, and perspectives into civilian communities and industries. By building connections with civilians who are interested in and supportive of the military, veterans can promote understanding, bridge cultural gaps, and foster collaboration. Relatable connections allow veterans to share their military expertise, contribute to civilian organizations, and create positive change by leveraging their skills and experiences in new contexts.

For veterans transitioning out of the military, establishing relatable connections to their service is of paramount importance. These connections provide a sense of belonging, understanding, and validation, while also offering peer support, mentorship, and opportunities for continued learning and growth. By staying connected to their military experiences through relatable connections, veterans can successfully navigate their transition, thrive in civilian life, and make meaningful contributions to their communities.

4. Fostering a Sense of Belonging: Transitioning from the military to civilian life can sometimes lead to a sense of isolation or a loss of identity for veterans. Overemphasizing their military identity may perpetuate this feeling and hinder their integration into civilian communities. By embracing a broader identity, veterans can actively seek out opportunities to connect with diverse groups of individuals, participate in nonmilitary activities, and engage in social and professional networks outside the military sphere. This fosters a sense of belonging, helps veterans feel valued beyond their military service, and supports their overall well-being as they navigate their postmilitary lives.

To navigate this balancing act, veterans can seek support from mentors, career counselors, and veterans' organizations that guide effectively presenting their skills and experiences in civilian contexts. They can also engage in professional-development opportunities, networking events, and workshops that expose them to new industries and perspectives. By actively cultivating a multifaceted professional identity and not overemphasizing their military background, veterans can maximize their potential, build meaningful connections, and thrive in their transition to civilian life.

Fostering a sense of belonging is crucial for veterans who are transitioning out of the military. The military community provides a unique and tightly knit support system, built on shared experiences, values, and a strong sense of camaraderie. As veterans transition into civilian life, they need to find a new sense of belonging to support their emotional well-being, social integration, and successful transition.

Nurturing a sense of belonging plays a pivotal role in fostering the emotional well-being of veterans. Military service often involves intense and challenging experiences that can have a lasting impact on veterans' mental and emotional health. Transitioning out of the military can be a complex and sometimes isolating process, leading to feelings of loss, disconnection, and uncertainty. By fostering a sense of belonging, veterans can find a supportive network of individuals who understand their experiences, share their values, and provide emotional support.

This sense of belonging helps alleviate feelings of isolation, reduces stress, and contributes to overall psychological well-being.

Transitioning from the military to civilian life can present social-integration challenges for veterans. The military provides a structured environment with clear roles and expectations, while the civilian world may be more diverse and less uniform in its social dynamics. Fostering a sense of belonging allows veterans to find communities, organizations, and social networks where they can connect with others who share common interests, goals, or backgrounds. These connections facilitate social integration; help veterans build new relationships; and provide opportunities for social support, networking, and shared activities. Feeling socially integrated enhances veterans' overall satisfaction and sense of belonging in their new civilian communities.

By fostering a sense of belonging, veterans can establish a robust support system beyond the military context. The military community is known for its strong bonds and support networks. Leaving this network behind can create a void in veterans' lives. By fostering a sense of belonging, veterans can build a new support system that provides encouragement, understanding, and guidance during their transition. This support system can consist of fellow veterans, friends, family, community organizations, or support groups specifically designed for veterans. Having a reliable support system in place helps veterans navigate challenges, seek advice, and find resources to support their overall well-being and successful transition.

The cultivation of a sense of belonging contributes to veterans' development of identity and purpose that extends beyond their military service. Military service often becomes a significant part of veterans' identity, and transitioning out of the military can sometimes lead to a loss of identity or a feeling of disconnection. By finding a new sense of belonging, veterans can explore different aspects of their identity, discover new passions and interests, and redefine their purpose in civilian life. This sense of belonging allows veterans to build a multifaceted identity that includes their military experiences but extends beyond them, providing a renewed sense of purpose and direction.

The sense of camaraderie and brotherhood experienced in the military is a powerful bond that fosters teamwork, loyalty, and mutual support. Fostering a sense of belonging allows veterans to recreate this bond to some extent in their civilian lives. Connecting with fellow veterans, joining veteran organizations or support groups, and participating in activities or events that honor their service cultivates a sense of camaraderie and brotherhood. This shared sense of belonging with others who have served builds a supportive network where veterans can find understanding, share stories, and create new memories, reinforcing their sense of identity and purpose.

When transitioning to civilian life, veterans may initially rely heavily on their existing military networks for job opportunities and professional connections. While these networks can provide valuable support and guidance, relying solely on them can limit the expansion of a veteran's civilian network. To overcome this limitation, veterans should actively seek opportunities to engage with professionals outside the military community. This can involve attending industry-specific networking events, participating in professional-development programs, joining relevant associations, and actively reaching out to civilians who share common interests or career aspirations. By broadening their networks and diversifying their connections, veterans increase their chances of discovering new opportunities and building relationships with professionals who may offer valuable insights and support.

Veterans, as individuals who have served their countries with honor and dedication, often find themselves facing unique challenges as they transition from military to civilian life. One crucial aspect of this transition is the need to establish and maintain robust networks, both personal and professional. While military networks can provide initial support, it is essential for veterans not to rely solely on these networks. Instead, veterans should strive to diversify their network usage by actively engaging with civilian networks.

Relying solely on military networks restricts exposure to a limited range of perspectives and experiences. By diversifying their networks, veterans open themselves up to a wealth of diverse perspectives, enabling them to gain valuable insights from individuals with different backgrounds, cultures, and professions. Engaging with civilians allows veterans to tap into a vast pool of knowledge and expertise that can broaden their horizons, challenge their preconceived notions, and foster personal growth. These diverse perspectives can also be invaluable when veterans embark on new career paths or seek innovative solutions to challenges in their personal lives.

While military networks can provide initial job leads and connections, relying solely on them can limit veterans' professional opportunities in the long run. Expanding their network to include civilian professionals allows veterans to tap into different industries, gain access to new job openings, and explore various career paths. Civilian networks often offer diverse skill sets and expertise, creating opportunities for veterans to acquire new skills or transition into civilian roles that align with their interests and passions. By diversifying their networks, veterans can access a broader range of resources, mentorship, and professional advice, ultimately increasing their chances of success in the civilian job market.

Transitioning from military to civilian life can be challenging and emotionally taxing. While military networks can provide a sense of camaraderie and understanding, veterans must establish a robust civilian support system as well. By diversifying their networks, veterans can forge meaningful relationships outside the military community, forming friendships based on shared interests, hobbies, and personal values. Building a diverse personal support system can provide veterans with emotional support, companionship, and a sense of belonging, all of which contribute to their overall well-being and successful transition to civilian life.

Diversifying networks expose veterans to different ways of thinking and problem-solving, enhancing their adaptability and resilience. Engaging with individuals from diverse backgrounds can help veterans develop a broader perspective and learn new approaches to overcoming challenges. By navigating through various social and professional environments, veterans acquire the skills necessary to adapt to unfamiliar situations and thrive in diverse settings. These enhanced adaptability and resilience skills are not only valuable for personal growth but also advantageous in the ever-evolving civilian job market.

While military networks undoubtedly offer a valuable support system, veterans should not rely solely on them when establishing and maintaining their networks. By diversifying their networks to include civilian connections, veterans gain access to a multitude of benefits, including diverse perspectives, expanded professional opportunities, stronger personal support systems, and enhanced adaptability. By actively engaging with civilian networks, veterans can ensure a smoother transition to civilian life, tap into new resources, and maximize their potential for personal and professional growth. It is through the diversification of networks that veterans can truly harness the full range of opportunities available to them beyond their military service.

5. Underestimating the Value of Civilian Experiences: Sometimes veterans may inadvertently limit their networking capabilities by underestimating the value of their civilian experiences. After spending years in the structured and hierarchical environment of the military, veterans may overlook or undervalue the skills and knowledge they gained before entering military service or during periods of civilian employment. This self-perception can limit their ability to effectively engage with civilian professionals and demonstrate their versatility and adaptability. Recognizing and highlighting the full range of their experiences, both military and civilian, allows veterans to showcase a more comprehensive skill set and enables them to connect with a broader range of professionals.

To address these common limitations, veterans must invest time in understanding the civilian networking landscape and adapting their approach accordingly. Seeking guidance from mentors, participating in networking workshops, and leveraging resources provided by veterans' organizations can all contribute to enhancing veterans' networking capabilities during their transition to civilian life. By translating their skills, finding the right balance in identity presentation, diversifying their networks, and recognizing the value of their civilian experiences, veterans can expand their networking horizons and tap into a wider range of opportunities in the civilian world.

Veterans may underestimate civilian experience because they perceive it as lacking the structure and discipline of the military. However, civilian experiences encompass a wide range of backgrounds, cultures, and industries, offering diverse perspectives that can broaden a veteran's worldview. Engaging with individuals who have not served in the military provides an opportunity to learn from different life experiences, approaches to problem-solving, and alternative perspectives. By embracing and valuing civilian experiences, veterans can gain new insights, challenge their assumptions, and foster personal growth.

Veterans often possess a wealth of skills that are highly transferrable to civilian careers, such as leadership, teamwork, adaptability, and resilience. However, they may overlook the fact that civilians also acquire valuable skills through their own experiences. Civilian experience offers a wide array of skills that can complement and enhance those gained in the military. For instance, civilians may excel in areas such as project management, sales, marketing, technology, or entrepreneurship, bringing a fresh perspective and complementary skill sets to the table. By recognizing the value of civilian experience, veterans can leverage these additional skills to diversify their professional profiles and enhance their career prospects.

Veterans often rely heavily on their military networks, which can result in underestimating the importance of networking and collaborating with individuals outside the military community. The civilian experience offers veterans the opportunity to build connections and establish professional relationships with individuals from various industries and backgrounds. Collaborating with civilians can lead to the discovery of new opportunities, mentorship, and the exchange of ideas that can significantly contribute to personal and professional growth. By underestimating the value of civilian networks, veterans may unintentionally limit their access to valuable resources and miss out on potential career advancement.

Civilian experiences can provide veterans with opportunities for personal growth and adaptability. Engaging with civilians allows veterans to embrace new perspectives, learn from different lifestyles, and develop a deeper understanding of diverse cultures. The ability to adapt to unfamiliar environments and connect with individuals from varying backgrounds is a valuable skill set that can enhance personal and professional success. By underestimating the value of civilian experience, veterans may miss out on the chance to broaden their horizons, develop empathy, and build a more inclusive worldview.

As veterans transition into civilian life, they must recognize the value of civilian experience. By embracing diverse perspectives, acknowledging transferable skills, and actively engaging with civilians, veterans can maximize their potential for personal and professional growth. Underestimating civilian experience not only limits their opportunities but also inhibits the development of a well-rounded perspective. By valuing civilian experience, veterans can enrich their knowledge and skills, forge new connections, and contribute to a more diverse and inclusive society. It is through the acknowledgment and appreciation of civilian experience that veterans can truly thrive in their postmilitary lives.

While military networks offer a sense of camaraderie and understanding, veterans often find it difficult to establish robust civilian support systems. Within our exploration, we examine the impact of veterans' constrained civilian networks on their mental well-being, with a specific focus on the emotions of isolation, lack of comprehension, and diminished availability of essential support resources. These factors have the potential to contribute to various mental health challenges experienced by veterans.

Veterans often encounter a lack of understanding and awareness of military culture and experiences within civilian networks. This lack of understanding can manifest in various ways, including misconceptions, stereotypes, or insensitive remarks. The resulting stigma can make veterans hesitant to disclose their military background or share their struggles, fearing judgment or being misunderstood. The inability to find acceptance and empathy within civilian networks can exacerbate feelings of isolation, create a sense of alienation, and further contribute to the deterioration of veterans' mental health.

Civilian networks play a vital role in connecting individuals with various support resources, including mental health services, counseling, and peer support groups. Limited civilian networks can hinder veterans' access to these crucial resources, leaving them without the necessary support to address their mental health concerns effectively. The lack of awareness or knowledge about available services specific to veterans may result in veterans struggling to find appropriate care tailored to their unique needs. This limited access to support resources can have severe consequences on veterans' mental health, prolonging their suffering and preventing timely intervention and treatment. Limited access to crucial support resources can have detrimental effects on the well-being and overall adjustment of veterans as they transition into civilian life. We will delve into the challenges faced by veterans, the consequences of reduced support, and potential strategies to address this issue.

Veterans often encounter difficulties in accessing support resources as they navigate the civilian landscape. One of the key challenges is the unfamiliarity with the civilian system and the lack of knowledge about available resources. This lack of understanding can hinder veterans from seeking the necessary support they need for their mental health.

Limited access to support resources can lead to profound feelings of isolation among veterans. Transitioning from a close-knit military community to civilian life, where connections may be more fragmented, can intensify this sense of isolation. Without a strong support network, veterans may experience a profound loss of camaraderie and understanding, exacerbating their feelings of loneliness and disconnection.

Reduced access to support resources can significantly impact the mental health of veterans. The absence of adequate support can lead to increased stress, anxiety, and depression, as veterans struggle to cope with the challenges of civilian life without the necessary assistance. Moreover, without appropriate resources and guidance, veterans may find it difficult to address and manage mental health issues effectively, leading to long-term consequences on their well-being.

To mitigate the adverse effects of reduced access to support resources, it is essential to implement strategies that bridge the gap between veterans and available resources. This includes raising awareness about support services specifically tailored for veterans, improving outreach efforts to ensure veterans are aware of the resources available to them, and enhancing collaboration between military and civilian organizations to create a seamless transition process.

The reduced access to support resources within civilian networks poses a significant challenge to veterans' mental health. The feelings of isolation, lack of understanding, and limited access to crucial support resources can contribute to mental health challenges among veterans. By recognizing and addressing this issue, we can work towards building stronger civilian networks that provide the necessary support and resources to enhance the overall well-being of veterans during their transition and beyond.

Transitioning from military to civilian life is a complex process that requires adjustment to new environments, routines, and social structures. Limited civilian networks can impede veterans' successful reintegration into society and hinder their ability to navigate this transition effectively. The absence of a supportive network can make it challenging for veterans to establish new social connections, find employment opportunities, or access relevant resources for their transition. This can contribute to feelings of frustration, helplessness, and low self-esteem, further deteriorating their mental well-being.

The reintegration of veterans into civilian life is a critical process that entails numerous challenges and opportunities. This page aims to explore the multifaceted impact of reintegration on veterans and the broader society. By understanding the various aspects of reintegration, we can identify strategies and initiatives to support veterans' successful transition, promote their well-being, and harness their unique skills and experiences.

Reintegration also has a significant impact on veterans' social connections and support networks. Transitioning from the tight-knit military community to civilian life can result in feelings of isolation and disconnection. The loss of camaraderie shared experiences, and structured support systems can lead to a sense of alienation. Building and maintaining social connections within civilian communities is vital to combatting isolation, fostering a sense of belonging, and providing much-needed support to veterans during their reintegration process.

One of the most critical aspects of reintegration is the successful transition to meaningful employment and economic stability. Veterans bring a unique skill set and experience to the civilian job market. However, translating their military skills and qualifications into the language and requirements of the civilian job market can be challenging. Unemployment or underemployment can lead to financial stress, diminished self-esteem, and a sense of purposelessness. Ensuring veterans have access to vocational training, career counseling, and employment opportunities that recognize and value their military skills is essential for their successful reintegration.

The process of reintegration has a profound impact on the dynamics and relationships within veterans' families. The strain of deployment and the challenges faced during military service can affect familial bonds and communication patterns. Reintegration requires a shared adjustment process, with both veterans and their families adapting to new roles and routines. Access to family counseling, support programs, and resources that address the unique needs of military families can strengthen these relationships and foster healthy family dynamics during the reintegration process.

The successful reintegration of veterans extends beyond individual well-being and encompasses their integration into the broader community. Veterans bring valuable skills, leadership qualities, and a strong work ethic that can contribute to the community's social fabric. Community engagement, volunteer opportunities, and initiatives that recognize and leverage veterans' strengths and experiences are essential for their meaningful integration into civilian life. By providing platforms for veterans to contribute and make a positive impact, communities can reap the benefits of their skills and dedication.

The impact of reintegration on veterans is far-reaching and complex. Addressing the psychological, social, economic, and familial aspects of reintegration is crucial in supporting veterans' successful transition into civilian life. By providing comprehensive support systems, tailored programs, and community integration opportunities, we can empower veterans to thrive, harness their unique talents, and contribute to the well-being and prosperity of both themselves and society as a whole.

Diversifying networks beyond the military community is essential for veterans' mental health and well-being. By actively engaging with civilian networks, veterans can expand their support systems, gain access to understanding individuals, and find a sense of belonging. Building connections with civilians who have knowledge and experience in mental health, community services, and peer support can provide veterans with the necessary resources and understanding they need to address their mental health challenges effectively.

By diversifying their networks, veterans can establish connections with individuals who come from diverse backgrounds, experiences, and perspectives. Engaging with a diverse range of people enables veterans to broaden their understanding of the world, challenge their preconceptions, and cultivate empathy. Exposure to diverse perspectives fosters critical thinking, creativity, and adaptability—essential qualities for navigating the complexities of civilian life.

Building diverse networks provides veterans with access to a wider range of support systems. In the military, camaraderie and a strong support structure are integral components of everyday life. Transitioning to civilian life can create a void in terms of social support. By diversifying networks, veterans can establish connections with individuals who share common interests, values, and goals. These new relationships can offer emotional support, guidance, and camaraderie, helping veterans navigate the challenges of reintegration.

Expanding networks enables veterans to explore a wider spectrum of career opportunities. In the civilian job market, personal connections and professional networks play a significant role in accessing job leads, mentorship, and professional development. By expanding their networks, veterans can tap into hidden job markets, gain exposure to different industries, and leverage connections for career advancement. Diverse networks can provide valuable insights, references, and recommendations that can enhance veterans' employment prospects and long-term career success.

Broadening networks nurtures personal growth and development among veterans. Exposure to individuals with diverse backgrounds and experiences broadens horizons and encourages continuous learning. Engaging with different perspectives cultivates open-mindedness, adaptability, and cultural competency. Veterans who actively seek out diverse networks often experience personal growth in areas such as communication skills, problem-solving abilities, and self-awareness. This growth extends beyond professional success and positively impacts various aspects of their lives.

Expanding networks is essential for facilitating veterans' social integration into civilian communities. Building connections with individuals from different backgrounds facilitates a sense of belonging, reduces feelings of isolation, and promotes a more inclusive society. Engaging with diverse networks enables veterans to participate actively in community activities, contribute to local initiatives, and promote cross-cultural understanding. By fostering social integration, veterans become valued members of their communities, forging bonds that enrich both their lives and the wider society.

The importance of diversifying networks for veterans cannot be overstated. By expanding their social connections beyond the military community, veterans gain access to new perspectives, support systems, career opportunities, and personal growth. Diverse networks contribute to veterans' successful reintegration into civilian life, enhance their overall well-being, and foster a more inclusive and understanding society. Embracing diversity in networks is a transformative step toward ensuring veterans' continued growth, resilience, and thriving in their postmilitary lives.

The limited civilian networks available to veterans can have detrimental effects on their mental health. Feelings of isolation, a lack of understanding, reduced access to support resources, and the impact on the transition process are all significant factors that contribute to veterans' mental health challenges. Veterans, their families, and society as a whole must recognize the importance of establishing diverse and inclusive civilian networks that provide understanding, support, and access to mental health resources. By addressing the limitations of civilian networks, we can better support veterans in their transition to civilian life, foster their mental well-being, and ensure they receive care and a sense of belonging.

Chapter 5

Mental Health and Well-Being

The mental health and well-being of veterans are of utmost importance as they navigate the challenges of transitioning from military to civilian life. Serving in the military exposes individuals to unique stressors and traumatic experiences that can have long-lasting effects on their mental health. This chapter aims to highlight the importance of mental health and well-being for veterans, emphasizing the implications on their overall quality of life, successful transition, personal relationships, and societal impact.

1. Post-Traumatic Stress and Trauma: Veterans often face a higher risk of experiencing PTSD and other trauma-related mental health conditions due to their exposure to combat, violence, or other traumatic events during their military service. Untreated PTSD and trauma can significantly impact veterans' daily functioning, relationships, and overall well-being. Prioritizing mental health and seeking appropriate support and treatment is essential for managing symptoms, reducing distress, and enhancing their quality of life.

PTSD is a prevalent mental health condition that affects many veterans after they leave military service. The experiences of combat, exposure to traumatic events, and the challenges of deployment can leave lasting psychological effects on individuals. This section aims to shed light on how PTSD continues to impact veterans even after they transition out of the military. PTSD often manifests through persistent and intrusive symptoms that persist beyond military service. Flashbacks, nightmares, and intrusive thoughts about past traumatic experiences can cause veterans to relive distressing memories. These symptoms can disrupt daily functioning, sleep patterns, and overall quality of life, making it challenging for veterans to reintegrate into civilian society. The persistent nature of these symptoms heightens stress levels and can lead to feelings of helplessness and isolation.

One of the core features of PTSD is the relentless reexperiencing of traumatic events. Individuals with PTSD often suffer from vivid and intrusive memories of the traumatic incident. These memories can manifest as flashbacks, wherein they feel as though they are reliving the traumatic event, with sensory impressions and intense emotions resurfacing. Such intrusive recollections can be triggered by various cues, including sights, sounds, smells, or even internal thoughts and emotions. The uncontrollable and distressing nature of these re-experiencing episodes further compounds the emotional toll on individuals with PTSD.

In addition to flashbacks, individuals with PTSD commonly experience intrusive thoughts and nightmares related to the traumatic event. Intrusive thoughts are recurrent and distressing mental images, often uninvited and intrusive, that intrude upon daily consciousness. These intrusive thoughts can be pervasive and intrusive, disrupting concentration, sleep, and overall cognitive functioning. Nightmares related to the traumatic event may be vivid, terrifying, and recurring, causing individuals to wake up in a state of fear and distress. The presence of such intrusive thoughts and nightmares perpetuates the emotional distress and psychological burden experienced by those with PTSD.

Hyperarousal and hypervigilance are common symptoms associated with PTSD. Individuals may find themselves in a state of heightened alertness, always on guard for potential threats. This constant state of hyperarousal and hypervigilance can lead to physical and emotional exhaustion, as individuals remain in a perpetual state of high stress and anxiety. Sleep disturbances, irritability, difficulty concentrating, and an exaggerated startle response are all signs of hyperarousal, which further diminishes the quality of life for individuals with PTSD.

To cope with the distressing nature of their symptoms, individuals with PTSD often engage in avoidance behaviors and experience emotional numbness. Avoidance may involve evading reminders of the traumatic event, such as avoiding certain places, people, or activities associated with the trauma. Individuals may also isolate themselves from social interactions or withdraw from relationships to prevent triggering memories or experiencing overwhelming emotions. Emotional numbness is another coping mechanism that manifests as a blunted affect, detachment from others, and an inability to experience pleasure or engage in once-enjoyed activities. These avoidance strategies and emotional numbing further isolate individuals and impede their ability to fully participate in life.

Persistent and intrusive symptoms of PTSD exert a significant toll on daily functioning and overall quality of life. The constant intrusion of traumatic memories, intrusive thoughts, nightmares, and hyperarousal hinders individuals' ability to concentrate, perform routine tasks, and engage in activities they once enjoyed. Relationships can be strained, as the emotional distance and avoidance behaviors impede meaningful connections and intimacy. The unrelenting presence of these symptoms also contributes to feelings of frustration, shame, guilt, and a sense of being permanently altered by the traumatic event, affecting self-esteem and overall psychological well-being.

PTSD not only affects individuals on an individual level but also has a significant impact on their relationships and social functioning. The symptoms associated with PTSD can strain interpersonal connections, impair communication, and hinder individuals' ability to engage in social activities.

The presence of PTSD can create a considerable burden on relationships, affecting interactions with family members, friends, and romantic partners. The symptoms of PTSD, such as hypervigilance, emotional numbing, and avoidance behaviors, can disrupt communication and emotional intimacy. Individuals with PTSD may struggle to express their emotions, leaving their loved ones feeling confused, frustrated, or disconnected. The impact of traumatic memories and flashbacks can cause individuals to withdraw or become emotionally distant, leading to feelings of isolation and strain within the relationship.

Individuals with PTSD may experience communication difficulties that can strain their relationships with loved ones. The emotional intensity and distress associated with PTSD symptoms can make it difficult for individuals to effectively communicate their needs, fears, and experiences. They may struggle to articulate their feelings or find it overwhelming to discuss traumatic memories. The resulting communication gaps can lead to misunderstandings, increased tension, and decreased emotional support within the relationship.

Individuals with PTSD may experience social withdrawal and isolation as a result of their symptoms. Avoidance behaviors, such as avoiding social gatherings, public places, or situations reminiscent of the traumatic event, can limit social interactions and participation in activities. The fear of triggers or overwhelming emotions can lead individuals to isolate themselves to protect themselves from potential distress. This withdrawal can lead to a sense of loneliness, exacerbate feelings of depression, and further isolate individuals from their support networks.

PTSD not only affects the individual with the disorder but also has a profound emotional impact on their loved ones. Witnessing the struggles and suffering of someone with PTSD can evoke a range of emotions, including frustration, helplessness, and compassion fatigue. Loved ones may struggle to understand the complexities of PTSD and the emotional turmoil experienced by their partner, family member, or friend. This emotional burden can strain relationships and require increased support and understanding from both parties.

The impact of community and social stigma can exacerbate the effects of PTSD, leading to additional challenges in relationships and social functioning. Misconceptions, stereotypes, and a lack of understanding about PTSD can lead to judgment, dismissiveness, or avoidance by others. This stigma can create additional barriers for individuals with PTSD to seek help, disclose their condition, or engage in social activities without fear of judgment or ridicule. The societal impact of stigma can contribute to feelings of shame, isolation, and a reluctance to seek the support necessary for healing and recovery.

The presence of PTSD significantly affects relationships and the ability to engage in social interactions. The strain on relationships, communication challenges, social withdrawal, and the emotional impact on loved ones all contribute to the complexities faced by individuals with PTSD. Recognizing the challenges and providing support, understanding, and empathy are crucial for fostering healthy relationships and social integration. Society must combat stigma, promote awareness, and provide resources to support both individuals with PTSD and their loved ones in navigating the challenges of this debilitating disorder. With compassion and support, individuals with PTSD can find solace, healing, and the opportunity to rebuild meaningful relationships and social connections.

PTSD often impairs veterans' occupational functioning, affecting their ability to secure and maintain employment. The intrusive thoughts, hyperarousal, and hypervigilance associated with PTSD can make it challenging for veterans to concentrate, meet job requirements, and adapt to work environments. Difficulties in managing stress, interacting with colleagues, and maintaining consistent attendance can hinder career advancement and job stability. The resulting occupational challenges can lead to financial strain, increased stress levels, and a sense of inadequacy.

Individuals with PTSD may experience difficulties in concentration and maintaining focus, impacting their productivity in the workplace. Intrusive thoughts, flashbacks, and hyperarousal can cause distractions, leading to difficulties in completing tasks and meeting deadlines. Individuals with PTSD may experience decreased attention span, memory problems, and reduced ability to multitask. These challenges can hinder job performance and may be perceived as a lack of competence or commitment by employers and colleagues.

Individuals with PTSD may have heightened sensitivity to triggers that remind them of their traumatic experiences. These triggers can be present in the work environment, such as loud noises, crowded spaces, or specific visual cues. Exposure to triggers can induce anxiety, panic attacks, or dissociation, leading to a decline in work performance and an increased risk of absenteeism or tardiness. The need to avoid triggers may limit job opportunities or prevent individuals from fully engaging in their work responsibilities.

PTSD can make it challenging for individuals to manage stress and cope with the pressures of the workplace. The constant state of hypervigilance and hyperarousal can make individuals more susceptible to stressors, even those unrelated to the traumatic event. The overwhelming nature of stress can exacerbate PTSD symptoms and contribute to emotional exhaustion. Individuals with PTSD may struggle to find healthy coping mechanisms, leading to increased levels of stress, burnout, and reduced job satisfaction.

The interpersonal challenges associated with PTSD can impact work relationships and teamwork. Difficulties in communication, emotional numbing, and avoidance behaviors can hinder effective collaboration and cooperation with colleagues. This can lead to misunderstandings, conflicts, and strained working relationships. The emotional instability and mood fluctuations that often accompany PTSD can make it difficult to regulate emotions in the workplace, further impacting professional interactions and contributing to a negative work environment.

PTSD can influence occupational choices and hinder career progression. Individuals with PTSD may avoid or struggle in occupations that involve high levels of stress, exposure to triggers, or a need for constant vigilance. This can limit job opportunities and prevent individuals from pursuing their desired career paths. The impact of PTSD on job stability and performance may also impede career progression, as individuals may struggle to meet performance expectations, cope with workplace demands, or pursue professional-development opportunities.

Occupational functioning can be significantly impacted by PTSD, presenting challenges in terms of job performance, career advancement, and job security. Impaired concentration and focus, heightened sensitivity to triggers, difficulties managing stress, interpersonal challenges, and limitations in occupational choice all contribute to the complex relationship between PTSD and the workplace. Employers and colleagues must foster a supportive and inclusive work environment, recognize the unique needs of individuals with PTSD, and provide accommodations and resources to help them thrive in their professional lives. By creating a supportive atmosphere, we can empower individuals with PTSD to navigate the workplace successfully, maximize their potential, and contribute positively to their chosen fields.

PTSD is frequently accompanied by co-occurring mental health issues, such as depression, anxiety disorders, and substance abuse. The presence of multiple conditions can compound the challenges veterans face after leaving the service. Co-occurring mental health issues further impact their overall well-being, increase the severity of symptoms, and hinder their ability to engage in daily activities and seek appropriate treatment. The self-medication tendencies associated with substance abuse can exacerbate the negative effects of PTSD and further impede recovery.

PTSD is often accompanied by an increased risk of substance-use disorders and other mental health issues. We will examine the underlying factors contributing to this connection, the impact on an individual's well-being, and the importance of integrated treatment approaches for addressing these co-occurring challenges.

Individuals with PTSD may turn to substance use as a means of self-medication to cope with the distressing symptoms of the disorder. The emotional pain, intrusive thoughts, and hyperarousal experienced in PTSD can be overwhelming, leading individuals to seek relief and escape through drugs or alcohol. Substance use may provide temporary respite from the debilitating symptoms of PTSD, numbing emotional distress, or help individuals manage anxiety and sleep disturbances. However, this self-medication strategy often leads to a cycle of dependence, worsening the symptoms of PTSD, and increasing the risk of developing a substance-use disorder.

PTSD is frequently associated with other mental health issues, such as depression, anxiety disorders, and substance-use disorders. The shared risk factors and common underlying mechanisms contribute to the co-occurrence of these conditions. Individuals with PTSD may experience a range of mental health challenges that are intertwined with their trauma-related symptoms. These co-occurring conditions can intensify the overall impact on an individual's psychological well-being, exacerbating symptoms and impairing daily functioning.

Biological and psychological factors play a role in the co-occurrence of substance use and other mental health issues in PTSD.

Neurobiological changes resulting from trauma and chronic stress can affect brain chemistry and increase vulnerability to both substance use and mental health disorders. Additionally, shared psychological mechanisms, such as avoidance, emotional dysregulation, and cognitive distortions, contribute to the development and maintenance of co-occurring conditions. These factors interact and reinforce each other, creating a complex web of interrelated challenges.

Substance use and co-occurring mental health issues in PTSD can mutually reinforce and escalate one another. Substance abuse may worsen the symptoms of PTSD, impairing emotional regulation, exacerbating anxiety and depression, and increase the frequency and intensity of intrusive thoughts and nightmares. On the other hand, the distressing symptoms of PTSD may fuel substance use as individuals attempt to self-medicate or numb their psychological pain. This bidirectional relationship creates a cycle of escalating symptoms and challenges, making it more difficult for individuals to break free from this harmful pattern.

Addressing both PTSD and co-occurring substance use and mental health issues requires integrated treatment approaches that take into account the complexity of these interrelated conditions. Comprehensive treatment plans may include evidence-based therapies such as cognitive-behavioral therapy (CBT), trauma-focused interventions, and motivational interviewing. It is crucial to address substance use disorders concurrently with PTSD, ensuring that individuals receive appropriate support, therapy, and pharmacological interventions when necessary. Coordinated care teams that specialize in trauma and dual diagnosis can provide integrated treatment, addressing the unique needs of individuals with co-occurring conditions.

The co-occurrence of substance use and other mental health issues in individuals with PTSD is a significant challenge that impacts overall well-being. Understanding the factors that contribute to this link and the mutual reinforcement of symptoms is essential for effective intervention and treatment. Integrated approaches that address PTSD, substance-use disorders, and co-occurring mental health issues can break the cycle of self-medication, promote healing, and enhance individuals' chances of recovery and long-term stability. By providing comprehensive and personalized care, individuals can regain control over their lives, improve their mental and physical health, and reintegrate into society with a renewed sense of purpose and hope for the future.

PTSD places veterans at a heightened risk of self-destructive behaviors and suicide. The emotional distress, feelings of hopelessness, and the burden of unprocessed trauma can lead to a sense of desperation. Without proper support and intervention, the risk of self-harm or suicidal ideation increases. It is crucial to recognize the warning signs, provide timely assistance, and ensure veterans have access to comprehensive mental health care to mitigate the risk of tragic outcomes.

Veterans often face traumatic experiences during their military service, which can have a lasting impact on their mental health. Exposure to combat, violence, and life-threatening situations can result in the development of PTSD. The burden of trauma, coupled with the challenges of transitioning to civilian life, can intensify feelings of isolation, hopelessness, and despair among veterans with PTSD, increasing the risk of suicide.

PTSD can significantly impair an individual's mental health, exacerbating symptoms such as intrusive thoughts, nightmares, hypervigilance, and emotional numbing. The relentless emotional distress and hyperarousal experienced by veterans with PTSD can overwhelm their coping mechanisms and erode their resilience. The constant struggle to manage these symptoms can leave veterans feeling helpless and trapped, contributing to suicidal ideation and an increased risk of self-harm.

PTSD is often accompanied by other mental health conditions, such as depression, anxiety disorders, and substance-use disorders. The presence of multiple mental health issues further increases the risk of suicide in veterans. These co-occurring conditions can exacerbate feelings of despair, heighten emotional instability, and diminish an individual's ability to cope with overwhelming thoughts and emotions. The combination of PTSD and comorbid mental health issues creates a complex web of challenges that contribute to the heightened risk of suicide.

Many veterans with PTSD experience social isolation and a sense of disconnection from their support networks. The stigma surrounding mental health issues may discourage veterans from seeking help or disclosing their struggles, further isolating them from potential sources of support. The lack of understanding and empathy from family, friends, and society at large can intensify feelings of loneliness, exacerbate symptoms, and increase the risk of suicide among veterans with PTSD.

Early intervention and support are crucial in mitigating the risk of suicide among veterans with PTSD. Timely identification, comprehensive assessment, and access to appropriate mental health services are essential for effective intervention. Providing veterans with a safe and supportive environment, along with access to evidence-based therapies and peer support groups, can help reduce feelings of isolation, promote healing, and enhance coping strategies. Additionally, educating communities and fostering an understanding of the unique challenges faced by veterans with PTSD is crucial for combating stigma and providing a supportive network.

PTSD significantly increases the risk of suicide among veterans, highlighting the urgent need for comprehensive support and intervention. The burden of trauma, its impact on mental health, co-occurring conditions, social isolation, and lack of support all contribute to the heightened risk. By prioritizing early intervention, destigmatizing mental health issues, and providing access to appropriate services and support, we can effectively address the risk of suicide among veterans with PTSD. It is our collective responsibility to honor their service by ensuring their well-being and providing the necessary resources to support their mental health and prevent tragic outcomes.

PTSD has a profound and lasting impact on veterans after they leave military service. The persistent and intrusive symptoms, impaired relationships and social functioning, occupational challenges, co-occurring mental health issues, and increased risk of self-destructive behaviors all highlight the urgent need for awareness, support, and effective treatment for veterans. By recognizing the enduring effects of PTSD, society can work toward creating a supportive environment that promotes understanding, destigmatizes seeking help, and provides accessible mental health resources for veterans. It is our collective responsibility to address the needs of veterans with PTSD and ensure their well-being, allowing them to heal, thrive, and reintegrate into civilian life successfully.

2. Successful Transition to Civilian Life: The transition from military to civilian life can be a complex and challenging process. Veterans may encounter difficulties in adapting to civilian routines, establishing new social connections, and finding meaningful employment. Mental health and well-being play a crucial role in facilitating a successful transition. Addressing mental health concerns allows veterans to effectively cope with the stressors of transitioning, manage emotional challenges, and maintain a positive outlook, thus enhancing their ability to navigate this critical phase of their lives.

Mental health and well-being significantly impact veterans' relationships, including those with their families, friends, and intimate partners. Untreated mental health conditions can lead to strained relationships, communication difficulties, and emotional withdrawal. Investing in mental health care and support not only benefits the individual veteran but also fosters healthier and more fulfilling relationships with their loved ones. By prioritizing mental health, veterans can develop effective coping strategies, communicate their needs, and maintain strong connections with those who are important to them.

Conditions affecting mental health can impede effective communication and emotional intimacy within personal relationships. Symptoms of PTSD, such as avoidance, emotional numbing, and hyperarousal, can make it difficult for veterans to express their emotions and connect with their loved ones on a deep level. They may struggle to communicate their experiences, needs, and feelings, leading to misunderstandings; strained communication; and a sense of emotional distance between partners, family members, and friends.

Challenges related to mental health can lead to substantial shifts in a veteran's behavior and mood, thereby affecting their relationships. Depression and anxiety may lead to withdrawal, irritability, and changes in sleep patterns and appetite. These shifts in behavior can disrupt routines and dynamics within relationships, causing frustration and confusion among loved ones. Partners, family members, and friends may struggle to understand and adapt to these changes, further straining the relationship.

The presence of mental health issues can contribute to increased conflict and tension in personal relationships. Symptoms of PTSD, such as anger, hypervigilance, and emotional reactivity, can lead to outbursts, arguments, and disagreements. The heightened stress levels and emotional instability associated with mental health challenges can strain the patience and understanding of partners, family members, and friends, leading to conflicts that may be difficult to resolve without appropriate support and communication strategies.

Mental health challenges may contribute to social isolation and relationship strain among veterans. The stigma surrounding mental health issues can make veterans hesitant to disclose their struggles or seek support, leading to feelings of loneliness and isolation. The impact of mental health on daily functioning and engagement in social activities may limit opportunities for veterans to maintain social connections and participate in shared interests, resulting in relationship strain and potential feelings of being disconnected from their loved ones.

Support and understanding from partners, family members, and friends play a crucial role in mitigating the impact of mental health challenges on personal relationships. Open and honest communication, empathy, and active listening can foster an environment of understanding and validation for veterans. Educating oneself about mental health conditions and their effects on relationships can also facilitate empathy and provide a foundation for effective support. Encouraging veterans to seek professional help and providing resources for mental health treatment and support networks can be instrumental in promoting their well-being and strengthening personal relationships.

The presence of mental health challenges among veterans can have a notable impact on various aspects of personal relationships, encompassing partnerships, family dynamics, and friendships. The difficulties in communication, changes in behavior and mood, increased conflict, and social isolation can strain these relationships. However, with support, understanding, and effective communication strategies, the negative effects of mental health challenges can be mitigated. By creating an environment of empathy, educating oneself, and providing access to mental health resources, we can foster healthier and more resilient personal relationships for veterans, enhancing their overall well-being and quality of life.

3. Enhanced Workforce Engagement: Employment is an essential aspect of veterans' reintegration into civilian life. Mental health and well-being have a direct impact on veterans' ability to engage in the workforce effectively. Addressing mental health concerns improves concentration, productivity, and overall job performance. It also enhances the ability to manage stress, adapt to new work environments, and maintain a healthy work-life balance. By prioritizing mental health and seeking appropriate support, veterans can maximize their potential in the workplace and contribute positively to their professional endeavors.

Veterans often possess a high level of resilience and adaptability, traits developed through their military experiences. Positive mental health among veterans enables them to effectively navigate and overcome challenges in the workplace. Their ability to remain calm under pressure, adapt to changing circumstances, and persevere through difficult situations enhances their problem-solving skills and resilience in the face of adversity. This resilience not only contributes to their success but also creates a positive work culture where others can learn from their example and develop their resilience.

Military service fosters strong leadership skills and a deep understanding of the importance of teamwork. When veterans have positive mental health, they can effectively utilize these skills in the civilian workforce. Their ability to lead by example, motivate others, and foster collaboration can significantly enhance team dynamics and organizational performance. Veterans often bring a strong sense of duty, discipline, and loyalty to their work, inspiring those around them and fostering a sense of camaraderie. Their experiences in diverse and multicultural environments during their service also contribute to their ability to navigate and appreciate diverse work settings.

Positive mental health among veterans enhances their problem-solving and decision-making abilities. Veterans are trained to analyze complex situations, make quick and informed decisions, and take calculated risks. These skills, coupled with their ability to think critically and adapt to changing circumstances, enable them to excel in various professional roles. Employers benefit from the innovative thinking and strategic mindset veterans bring to the table, as they offer unique perspectives and creative solutions to workplace challenges.

Veterans with positive mental health often demonstrate a strong work ethic and a commitment to excellence. Their dedication, discipline, and attention to detail translate into high-quality work and a commitment to meeting deadlines and goals. Veterans understand the value of teamwork and cooperation, making them reliable and trustworthy employees. Their strong sense of responsibility and accountability contributes to a positive work environment, where others are motivated to perform at their best.

Fostering positive mental health among veterans promotes higher employee engagement and contributes to a positive organizational culture. When veterans feel supported, valued, and included in the workplace, their sense of belonging and overall job satisfaction increases. This, in turn, leads to higher levels of employee engagement, productivity, and retention. By creating a work environment that recognizes and supports the mental well-being of veterans, organizations can foster a culture of inclusivity, resilience, and commitment, benefiting both the veterans themselves and the entire workforce.

Positive mental health among veterans has a significant impact on the workforce and employee engagement. Their resilience, adaptability, leadership skills, problem-solving abilities, commitment, and work ethic bring valuable contributions to the workplace. Organizations that prioritize the mental well-being of veterans create an environment that values their unique strengths and experiences, fostering a positive work culture that benefits all employees. By recognizing and supporting the positive mental health of veterans, organizations can harness their potential and contribute to a more engaged, productive, and inclusive workforce.

4. Social and Societal Implications: The mental health and well-being of veterans have broader social and societal implications. When veterans' mental health needs are met, they are more likely to engage in their communities, contribute to society, and advocate for important causes. Conversely, untreated mental health conditions can lead to social withdrawal, substance abuse, homelessness, and involvement in the criminal justice system. By prioritizing mental health and well-being, veterans can become empowered individuals who actively participate in society and inspire positive change.

The societal stigma surrounding mental health often leads to veterans feeling isolated and misunderstood. The perception that mental health struggles are a sign of weakness or a lack of resilience can prevent veterans from seeking help and support. This social stigma creates barriers to open communication and can exacerbate feelings of isolation and alienation among veterans. The societal expectation that veterans should be self-reliant and unaffected by their experiences can compound the challenges they face in accessing appropriate mental health care and reintegrating into their communities.

Veterans with mental health challenges often face difficulties in finding and maintaining employment, which has broader societal consequences. The symptoms associated with mental health conditions can affect job performance, attendance, and productivity. Discrimination and lack of understanding from employers may further hinder veterans' ability to secure suitable employment opportunities. The economic impact of unemployment or underemployment among veterans can lead to financial instability and dependence on social-support systems, placing an additional burden on society as a whole.

Veterans' mental health challenges place a strain on health-care systems, including both veteran-specific services and general mental health resources. The demand for mental health support often exceeds the capacity of existing services, leading to long wait times and limited access to care. Insufficient resources can result in delayed or inadequate treatment, contributing to prolonged suffering and diminished quality of life for veterans. Addressing the mental health needs of veterans requires increased investment in mental-health-care infrastructure, training of mental health professionals, and the development of specialized programs tailored to the unique challenges faced by veterans.

The mental health challenges experienced by veterans also affect their families and relationships, with societal implications. The emotional and psychological toll of mental health conditions can strain family relationships, leading to increased stress, conflict, and breakdowns in communication. This ripple effect extends to children, spouses, and extended family members, impacting their well-being and overall functioning. The societal cost of supporting and addressing the consequences of strained familial and social relationships further underscores the need for comprehensive mental health care and support services.

The successful integration and reintegration of veterans into society can be hindered by mental health challenges. The symptoms associated with mental health conditions may make it difficult for veterans to participate in community activities, engage in social interactions, and establish support networks. This social disconnection can contribute to a sense of isolation and hinder the overall well-being and satisfaction of veterans within society. The societal cost of limited social integration includes missed opportunities for veterans to contribute their skills and experiences to their communities, limiting the potential for their positive impact on society.

The mental health challenges faced by veterans have significant societal implications, ranging from social stigma and isolation to employment difficulties, strain on health-care systems, and impacts on families and relationships. Addressing these challenges requires a comprehensive approach that includes destigmatizing mental health, providing accessible and appropriate mental-health-care services, and promoting social integration and support networks. By acknowledging and addressing the societal impact of veterans' mental health challenges, society can better support veterans in their journey to recovery, resilience, and successful reintegration into civilian life.

The mental health and well-being of veterans are essential for their overall quality of life, successful transition, personal relationships, and societal impact. Addressing mental health concerns, such as post-traumatic stress and trauma, plays a significant role in facilitating a successful transition to civilian life, maintaining healthy relationships, and maximizing workforce engagement. By prioritizing mental health and seeking appropriate support and treatment, veterans can lead fulfilling lives, contribute to their communities, and serve as role models for resilience and growth. Society must recognize and support the mental health needs of veterans, ensuring their well-being and providing a foundation for their continued success and fulfillment.

Veterans with mental health challenges often face difficulties in finding and maintaining employment, which can have a significant impact on their financial stability. Symptoms associated with mental health conditions, such as difficulties concentrating, mood swings, and anxiety, can affect job performance, attendance, and productivity. This may lead to increased absenteeism, reduced work output, and a higher risk of unemployment or underemployment. The resulting loss of income and limited career advancement opportunities can undermine veterans' financial stability and make it difficult to meet financial obligations and achieve long-term financial goals.

Mental health challenges among veterans can result in increased health-care costs, further straining financial stability. Seeking appropriate mental health care, medications, and therapy can be expensive, particularly for veterans without adequate health insurance coverage. Additionally, co-pays, deductibles, and out-of-pocket expenses can quickly accumulate, placing an additional burden on veterans' financial resources. The financial strain of accessing necessary mental health treatment can limit veterans' ability to allocate funds toward other essential expenses, such as housing, education, and retirement planning.

Veterans' financial decision-making abilities can be compromised by mental health challenges, resulting in unfavorable financial choices that may further contribute to financial instability. Symptoms like impulsivity, emotional instability, and difficulties with concentration and memory can impair veterans' ability to make sound financial judgments. This may result in impulsive spending, high-risk investments, excessive debt, and poor financial planning, all of which can negatively impact their financial stability over time. Addressing mental health challenges and providing resources for financial literacy and management can help veterans improve their financial decision-making skills and mitigate the risks associated with impaired judgment.

Veterans' mental health challenges can influence their relationship with debt and credit, further impacting their financial stability. High levels of stress, anxiety, and depression can lead to impulsive spending, self-medication through retail therapy, or the accumulation of consumer debt. Additionally, financial challenges resulting from employment difficulties or health-care costs can force veterans to rely on credit cards or loans to meet their financial obligations. The resulting debt burden can lead to financial stress, limited access to credit in the future, and hindered financial mobility and stability.

Addressing veterans' mental health is crucial for promoting financial stability. Accessible and comprehensive mental health support services can help veterans manage their mental health challenges effectively, enhancing their ability to obtain and maintain stable employment, make sound financial decisions, and reduce financial stress. Interventions that provide mental health treatment, financial counseling, and employment assistance tailored to veterans' unique needs can significantly improve their financial well-being and overall quality of life.

The mental health of veterans is a critical factor in their financial stability. Employment challenges, increased health-care costs, impaired financial decision-making, and the relationship between debt and credit can all contribute to financial instability among veterans with mental health challenges. Recognizing the importance of mental health support and providing resources to address mental health concerns can help veterans overcome these challenges, enhance their financial well-being, and promote long-term financial stability. By prioritizing veterans' mental health and supporting their financial literacy and management, society can help ensure that veterans can support their families and goal for future endeavors.

Chapter 6

Financial Stability

As veterans transition from military service to civilian life, achieving and maintaining financial stability is crucial for their successful reintegration. Financial stability provides a solid foundation that enables veterans to meet their basic needs, pursue further education or career opportunities, and establish a sense of security for themselves and their families. This chapter explores the significance of financial stability for veterans during their transition, emphasizing the benefits it brings and the challenges they may encounter without it.

1. Meeting Basic Needs: Financial stability is essential for veterans to meet their basic needs, such as housing, food, health care, and transportation. A stable income ensures access to safe and affordable housing, adequate nutrition, and quality health care. These fundamental aspects of well-being lay the groundwork for veterans to focus on other aspects of their transition, such as education, employment, and building meaningful relationships. Without financial stability, veterans may face difficulties in accessing these necessities, which can hinder their overall well-being and success during the transition period.

One of the primary challenges veterans face after leaving the service is securing stable and affordable housing. Factors such as sudden relocation, limited financial resources, and difficulties in accessing housing assistance programs contribute to this challenge. Some veterans may experience homelessness or precarious living situations, which can have profound impacts on their overall well-being. Supportive housing programs specifically designed for veterans, such as the US Department of Veterans Affairs (VA) Supportive Housing (HUD-VASH) program, provide vital resources, including rental assistance and case management, to help veterans secure stable housing and address their unique needs.

Another VA-supported program is the Veterans Health Administration's Grant and Per Diem (GPD) program, which provides funding to community-based organizations that offer transitional housing, treatment services, and support for homeless veterans. GPD programs focus on providing temporary housing and facilitating veterans' transition to permanent housing while addressing their specific challenges such as substance abuse or mental health issues.

The VA also administers home-loan programs to help veterans become homeowners. The VA Home Loan program provides opportunities for eligible veterans to secure favorable mortgage terms and access to affordable housing. Veterans can obtain VA-backed loans with reduced down payments, competitive interest rates, and no private mortgage insurance requirement. Additionally, the VA offers grants to adapt housing for veterans with service-connected disabilities, promoting accessibility and independent living.

Community-based organizations and nonprofit agencies play a crucial role in providing housing support to veterans. These organizations often collaborate with the VA and local authorities to offer transitional housing, rental assistance, and other services. For example, organizations like Habitat for Humanity and Purple Heart Homes provide housing solutions, including home repairs, renovations, and new-home construction, specifically for veterans. These initiatives focus on improving veterans' housing conditions and promoting long-term stability.

Many states and local governments have implemented housing initiatives tailored to veterans' needs. These programs can include affordable housing projects, rental assistance, and partnerships with local landlords to ensure veterans have access to safe and affordable housing options. State-level Veterans Affairs departments often provide information and resources on housing programs available within their jurisdiction, helping veterans navigate the various options and eligibility criteria.

In addition to housing programs, supportive services are crucial for veterans' successful housing transition. Case management, counseling, and job-placement services can assist veterans in overcoming barriers to housing stability. Organizations such as Veterans Affairs Supportive Housing (VASH) case managers, Local Veterans Employment Representatives (LVERs), and Disabled Veterans Outreach Program (DVOP) specialists work with veterans to address their individual needs and connect them with resources for securing housing and employment.

Securing stable housing is a vital step for veterans as they transition from military service to civilian life. The VA, community-based organizations, state and local initiatives, and supportive services offer various housing options and resources to meet veterans' specific needs. Through programs like HUD-VASH, VA-backed home loans, community-housing initiatives, and supportive services, veterans can access safe and affordable housing, receive necessary support services, and establish stable housing as they reintegrate into civilian communities. Continued collaboration and support from governmental and nonprofit organizations are crucial to ensuring that veterans have access to the housing opportunities they deserve and need to thrive in their postservice lives.

Food insecurity is another pressing issue faced by veterans during their transition. Economic challenges, limited income, and inadequate access to healthy food options can make it difficult for veterans to meet their nutritional needs. Some veterans may rely on food assistance programs, such as the Supplemental Nutrition Assistance Program (SNAP) or local food banks, to mitigate food insecurity. Organizations like the Veterans' Pantry and Feeding America also provide targeted support to veterans, ensuring they have access to nutritious meals. Addressing food insecurity among veterans is crucial to their overall well-being and can contribute to their successful reintegration into civilian life.

Food insecurity, the lack of consistent access to nutritious and sufficient food, is a significant issue affecting many individuals and households, including veterans. Despite their service to the nation, veterans face unique challenges that can contribute to food insecurity during and after their military service. Several factors contribute to food insecurity among veterans. One factor to consider while transitioning from military service to civilian life involves periods of unemployment, underemployment, or financial instability. Veterans may face difficulties in finding suitable employment, translating military skills to the civilian job market, or adjusting to new career paths, which can lead to limited financial resources and strain their ability to afford nutritious food consistently. Additionally, veterans with service-related injuries or disabilities may experience physical or mental health limitations that affect their ability to work or engage in income-generating activities, exacerbating financial constraints and making it challenging to afford an adequate food supply.

Veterans residing in food deserts or areas with limited access to affordable and healthy food options may struggle to obtain nutritious meals due to factors such as lack of transportation, limited availability of grocery stores or farmers' markets, and high food costs. Lastly, some veterans may be unaware of the support services and resources available to them, such as food assistance programs, community food banks, or veteran-specific organizations that provide food aid, leading to a lack of access to the help they need.

Food insecurity has detrimental effects on the overall well-being of veterans. Inadequate access to nutritious food can lead to various health consequences, including nutrient deficiencies, chronic diseases, and compromised immune systems. This is particularly significant for veterans who may already face heightened health risks due to service-related injuries or conditions, making proper nutrition essential for their overall well-being and recovery.

Food insecurity can have negative impacts on veterans' mental and emotional health. The uncertainty and stress associated with not knowing when or where the next meal will come from can contribute to feelings of anxiety, shame, and stigma. This can further exacerbate existing mental health conditions such as PTSD or depression, posing additional challenges to veterans' well-being.

Food insecurity also affects not only veterans but also their families. Insufficient food availability can strain family relationships, increase parental stress, and have detrimental effects on the overall well-being of children. Proper nutrition is crucial for children's development, academic performance, and long-term health; and the lack thereof can hinder their well-being when they do not have access to adequate food.

Additionally, food insecurity poses challenges to veterans' successful reintegration into civilian life. It creates additional barriers and stressors that impede their ability to focus on employment, education, and building a stable future. Addressing food insecurity is therefore crucial in supporting veterans' overall well-being and facilitating their successful transition to civilian life. By ensuring access to nutritious food, we can help alleviate some of the burdens veterans face and provide a solid foundation for their well-being and future endeavors.

Efforts to address food insecurity among veterans require a multifaceted approach:

a. Outreach and Education: Raising awareness about available resources is essential. Government agencies, nonprofit organizations, and community groups can collaborate to ensure that veterans are informed about food assistance programs, community food banks, and other local initiatives that provide support.

b. Enhancing Access: Improving access to affordable, nutritious food is crucial. Expanding the availability of farmers' markets, mobile food pantries, and transportation options in underserved areas can help veterans overcome geographical barriers.

Access to quality health care is a vital need for veterans, particularly as they navigate the transition from military health-care systems to civilian health-care providers. Many veterans face challenges in accessing affordable health care, including gaps in insurance coverage, long wait times for appointments, and difficulties in navigating the complex health-care system. The VA plays a crucial role in providing comprehensive health-care services to eligible veterans, but it may not cover all their needs. Expanding access to affordable health care, improving coordination between the VA and private health-care providers, and increasing awareness of available resources can support veterans in meeting their health-care needs.

Access to quality health care is a fundamental right, especially for those who have served their country. Veterans, having made sacrifices in the line of duty, deserve comprehensive health-care services that address their unique physical and mental health needs. Veterans face specific challenges when it comes to accessing health-care services. One of the primary challenges is the transition from military health-care systems to civilian providers. Navigating the civilian health-care landscape, transferring medical records, and finding providers who understand their specific health needs can be complex and daunting for veterans.

Geographic barriers also pose challenges for veterans, particularly those residing in rural or remote areas. Limited access to health-care facilities, specialists, and comprehensive medical services can result in delays in care, increased travel burdens, and difficulties accessing specialized treatments. These barriers can significantly impact veterans' ability to receive timely and appropriate health care.

Financial barriers can also hinder veterans' access to health-care services. While the VA provides health-care services to eligible veterans, some may encounter financial challenges. Veterans without VA health-care coverage or those with limited income may struggle with out-of-pocket costs for medications, treatments, or procedures not covered by their insurance. These financial barriers can limit veterans' ability to receive the necessary care they require.

Veterans often require specialized mental health services to address conditions such as PTSD, depression, and anxiety. However, there may be a shortage of mental health providers trained in veteran-specific issues, resulting in limited access to appropriate care. This shortage of mental health professionals who understand the unique needs of veterans further compounds the challenges veterans face in accessing adequate mental health care.

Addressing these challenges and improving access to health-care services for veterans is crucial. It requires efforts to streamline the transition process, expand health-care facilities in rural areas, address financial barriers, and enhance the availability of specialized mental health services. By ensuring that veterans can readily access the health-care services they need, we can support their overall well-being and honor their sacrifices in service to their country.

The Department of Veterans Affairs (VA) plays a critical role as a primary provider of health-care services for veterans, offering a comprehensive range of medical, surgical, and mental health services. The VA operates an extensive network of hospitals, clinics, and community-based outpatient centers throughout the country, ensuring accessible health care for veterans.

One key aspect of the VA health-care system is eligibility and enrollment. Eligible veterans can enroll in the VA health-care system, gaining access to a wide array of services, including preventive care, primary care, specialty care, and medications. Enrollment criteria are based on factors such as service-connected disabilities, income thresholds, and other qualifying criteria, ensuring that veterans receive the care they need.

The VA is uniquely equipped to address the specialized health care needs of veterans. It provides services specifically tailored to conditions related to military service, such as traumatic brain injury, exposure to environmental hazards, and combat-related injuries. Additionally, the VA prioritizes mental health services, offering a range of programs and therapies designed to address the unique mental health challenges veterans may face.

Care coordination is a significant focus within the VA health-care system. The VA emphasizes the importance of coordinating care to ensure veterans receive comprehensive and integrated health care. This involves case management, referral services, and collaboration among different health-care providers within the VA system. Care coordination efforts aim to ensure continuity of care, prevent gaps in treatment, and enhance the overall health-care experience for veterans.

The VA has also embraced telehealth services as a means to enhance access to care for veterans. Telehealth enables veterans to consult with health-care providers remotely, reducing travel burdens and improving access to specialists. This technology-driven approach enhances convenience, especially for veterans residing in remote or underserved areas, and expands the reach of health-care services.

The VA's extensive health-care system, specialized services, emphasis on care coordination, and adoption of telehealth services collectively contribute to the comprehensive and accessible health care available to veterans. By continually improving and expanding these services, the

VA strives to meet the evolving health-care needs of veterans and honor their service to the nation.

Ensuring veterans have access to high-quality health care requires collaboration between the VA, community health-care providers, and nonprofit organizations. Partnerships with local hospitals and medical centers can expand access to specialized services and reduce wait times. Collaborations with mental health organizations, universities, and research institutions can promote innovation and knowledge sharing to address veterans' mental health challenges effectively.

Transportation is a fundamental need that impacts veterans' ability to access employment, health care, education, and social activities. Lack of reliable transportation options, especially in rural or underserved areas, can limit veterans' mobility and independence. Public transportation subsidies; ride-sharing programs; and community volunteer networks, such as the Disabled American Veterans (DAV) Transportation Network, offer vital support to veterans in meeting their transportation needs. These programs facilitate access to essential services and promote veterans' overall well-being and engagement in their communities.

Access to reliable transportation is essential for veterans to maintain their independence, access health-care services, seek employment opportunities, and engage in community activities. Veterans often face unique challenges related to transportation, such as limited mobility due to service-related disabilities, lack of transportation options in rural areas, or financial constraints.

Transportation plays a crucial role in veterans' ability to access health-care services. Many veterans rely on the VA for their medical needs; and transportation is often necessary to reach VA facilities, clinics, or medical appointments. Lack of accessible transportation can result in missed or delayed appointments, hindering veterans' access to critical health-care services and potentially impacting their health outcomes.

Access to transportation is crucial for veterans who are pursuing employment or vocational opportunities. Access to reliable transportation enables veterans to commute to work, attend job interviews, and participate in training programs or educational opportunities. Having reliable transportation can expand employment options, enhance career mobility, and promote financial stability for veterans transitioning into civilian life.

Transportation is a key factor in facilitating veterans' social engagement and community integration. It allows veterans to participate in social activities, attend support group meetings, or engage in recreational and cultural events. Lack of transportation can lead to social isolation, limiting veterans' ability to connect with their peers, engage in community activities, and build a support network, which is essential for their overall well-being and sense of belonging.

Veterans residing in rural or remote areas often face challenges related to transportation. Limited public transportation options, long distances to essential services, and the lack of infrastructure can isolate veterans, making it difficult to access health care, employment, or other necessary resources. Establishing transportation solutions tailored to the specific needs of veterans in rural areas is crucial to ensure their access to essential services and promote community connectivity.

Many veterans face mobility challenges due to service-related disabilities or injuries. Accessibility considerations are essential to ensure that transportation options cater to veterans with physical limitations, such as wheelchair-accessible vehicles, ramps, and specialized transportation services. Adapting transportation infrastructure and vehicles can improve veterans' mobility and independence, empowering them to participate fully in community life.

Addressing transportation barriers for veterans requires collaborative efforts among government agencies, nonprofit organizations, and local communities. The VA, in collaboration with transportation authorities and service providers, can develop programs that offer discounted fares, specialized transportation services, or partnerships with ridesharing companies to ensure veterans have reliable transportation options. Nonprofit organizations, veteran service organizations, and community groups can establish volunteer-driven transportation services to bridge gaps in transportation access.

Advancements in technology, such as ride-sharing apps and telehealth services, can play a significant role in improving veterans' transportation access. Telehealth services reduce the need for in-person appointments, offering convenience and accessibility, especially for veterans in remote areas. Ride-sharing apps can provide flexible transportation options, reducing reliance on personal vehicles and offering cost-effective solutions for veterans without access to private transportation.

Ensuring veterans have access to reliable transportation is crucial for their well-being, health-care access, employment opportunities, and community integration. Addressing transportation barriers requires collaborative efforts, supportive initiatives, and the adoption of innovative solutions. By prioritizing transportation access, we can empower veterans to lead independent and fulfilling lives, facilitate their integration into civilian communities, and honor their service by ensuring they can access the resources and support they need to thrive beyond their military service, fostering a strong sense of belonging and well-deserved recognition in society.

The availability of support systems and resources is critical for veterans to meet their basic needs successfully. Nonprofit organizations, government agencies, and community-based initiatives play a significant role in providing support and resources tailored to veterans' unique needs. These resources include case management services, emergency financial assistance, employment assistance programs, and educational opportunities. Collaboration between public and private entities is essential to ensure comprehensive support for veterans, empowering them to overcome challenges, and establish stability in their postservice lives.

Meeting basic needs is a fundamental challenge faced by veterans after their military service. Housing, food, health care, and transportation are crucial pillars that support veterans' successful reintegration into civilian life. Recognizing and addressing the challenges veterans encounter in meeting these needs is essential to ensure their well-being and overall success. By providing targeted support systems, resources, and collaborations between government agencies and community organizations, we can bridge the gap between the services available and the actual needs of veterans, creating a safety net that enables them to rebuild their lives with dignity and purpose, thereby honoring their sacrifice and dedication to our country.

2. Pursuing Education and Training: Financial stability plays a critical role in enabling veterans to pursue further education and training. Many veterans seek additional education or training to enhance their skills and qualifications, making them more competitive in the civilian job market. Access to financial resources, such as tuition assistance, scholarships, or grants, can support their educational pursuits. Financial stability provides veterans with the means to cover educational expenses, including tuition, books, and living costs, reducing financial stress and allowing them to focus on their studies and professional development.

Transitioning from military service to civilian life can involve periods of unemployment or income disruption. Veterans may face challenges in securing stable employment immediately after leaving the service, leading to financial uncertainty and limited resources to invest in education. The rising cost of higher education presents a significant barrier for veterans. Tuition fees, textbooks, housing, and other related expenses can pose financial burdens, especially for veterans who do not qualify for full tuition coverage under the GI Bill or other educational assistance programs. Pursuing education often requires veterans to forgo employment or reduce working hours, leading to a loss of income. This trade-off between income and education can impact their financial stability and create additional financial strain.

Veterans who have dependents, service-related disabilities, or other financial responsibilities may face greater challenges in allocating resources for education. They may need to prioritize immediate financial needs over long-term educational investments, affecting their ability to pursue education without adequate financial support.

Financial stability plays a vital role in enabling veterans to access higher education opportunities. By providing financial support, such as scholarships, grants, and student loans, veterans can afford tuition fees, course materials, and living expenses, reducing financial barriers to education. Institutions and government programs that offer discounted tuition rates or fee waivers for veterans can significantly reduce the financial burden, contributing to the affordability and cost reduction of education. Additionally, financial-aid packages, GI Bill benefits, and other educational assistance programs help bridge the affordability gap, making education more accessible and affordable for veterans.

Financial stability allows veterans to focus on their educational pursuits without excessive financial worries. By alleviating financial stress, veterans can dedicate their time and energy to their studies, enhancing their chances of academic success and graduation. Education can enhance veterans' employability and provide them with the necessary skills and qualifications for career advancement. By investing in education, veterans can expand their job prospects, secure higher-paying positions, and achieve greater financial stability in the long run. Overall, fostering financial stability among veterans not only supports their educational aspirations but also opens doors to posteducation employment opportunities, ultimately contributing to their long-term financial well-being.

It is essential to raise awareness among veterans about the financial resources available to support their education. This includes informing them about the GI Bill, scholarships, grants, and other financial aid programs specific to veterans. Collaborative efforts between educational institutions, government agencies, and veteran service organizations are crucial to streamline and enhance the financial support system for veterans. This can involve establishing dedicated veteran support offices, improving outreach efforts, and simplifying the application process for financial aid.

In addition, providing comprehensive financial counseling services tailored to veterans' unique needs can empower them to make informed decisions about educational investments, budgeting, and managing their financial resources effectively. By offering financial counseling, veterans can gain valuable insights and guidance to navigate the complexities of financing their education and plan for long-term financial stability.

Advocating for policies that address the financial barriers faced by veterans in pursuing education is essential. Public policy initiatives can include increasing funding for educational assistance programs, expanding eligibility criteria, and supporting legislation that promotes affordable education for veterans. By actively engaging in public-policy discussions and advocating for change, stakeholders can work toward a more inclusive and supportive environment for veterans seeking higher education.

3. Employment and Career Advancement: Financial stability is closely tied to employment and career advancement for veterans. A stable income from employment allows veterans to support themselves and their families, reducing financial uncertainty. Financial stability provides the opportunity to save for future goals, invest in professional development, and plan for retirement. With a stable financial foundation, veterans can make informed career choices based on their interests and long-term goals, rather than solely on immediate financial needs. Financial stability also facilitates the transition to civilian employment, as it provides a buffer during the often-challenging process of job searching and initial adjustment.

Unemployment and underemployment are common challenges that veterans may face when transitioning to civilian life. They may experience difficulties in securing employment that aligns with their skills, qualifications, and career aspirations. Factors such as limited civilian-work experience, translating military skills to the civilian job market, and biases or misconceptions about veterans' abilities can contribute to higher rates of unemployment or underemployment. These obstacles hinder veterans from fully utilizing their talents and finding fulfilling career opportunities.

The lack of financial stability adds significant pressure to veterans, forcing them to prioritize immediate income over long-term career goals. Financial constraints may lead them to accept low-paying or unstable jobs that do not utilize their full potential or provide growth opportunities. This compromises their ability to build a sustainable and prosperous career, hindering their overall financial well-being and stability.

Some veterans may encounter challenges in obtaining civilian credentials or licenses necessary for certain professions. Variations in licensing requirements, difficulties in translating military training and certifications, or lack of awareness about available resources to support the credentialing process can impede veterans' ability to pursue desired careers. This credentialing barrier limits their access to specific occupations and professional advancement, further exacerbating the challenges they face in the civilian job market.

Addressing these issues requires targeted support and resources to bridge the gap between military experience and civilian employment, provide financial stability, and facilitate the credentialing process. By recognizing and addressing these barriers, society can better support veterans in their transition to civilian careers and help them achieve long-term success and financial security.

Financial stability plays a crucial role in several aspects of veterans' career journeys. Firstly, it enables them to navigate the job search process effectively, including traveling for interviews, attending networking events, or relocating to pursue employment opportunities. Having the necessary financial resources can enhance their chances of finding suitable employment and exploring various career options.

Veterans can leverage financial stability to invest in their ongoing education, professional development, and the enhancement of their skills. By having access to resources for obtaining certifications, attending workshops, or pursuing advanced degrees, veterans can equip themselves with the necessary qualifications and knowledge for career advancement.

In addition, financial stability enables veterans to participate in networking events, professional associations, and mentorship programs. Building connections and receiving guidance from experienced professionals can open doors to employment opportunities, career advice, and long-term professional relationships, thus enhancing their career prospects.

Financial stability can empower veterans to pursue entrepreneurial ventures and start their businesses. Having access to financial resources and support increases their chances of success, promotes self-employment, and contributes to economic growth in their communities.

Financial stability plays a pivotal role in various aspects of veterans' career journeys. It allows them to navigate the job search process, invest in education and skill enhancement, participate in networking and mentoring opportunities, and pursue entrepreneurial ventures. By fostering financial stability, society can support veterans in achieving their career aspirations, contributing to their long-term professional success and overall well-being.

Developing specialized job-placement programs that cater to veterans' unique skills and experiences can facilitate their integration into the civilian workforce. These programs can offer tailored job-matching services, assistance with resume-building, interview preparation, and mentorship opportunities.

Providing comprehensive financial education and counseling services to veterans equips them with the knowledge and skills necessary to manage their finances effectively. This includes guidance on budgeting, debt management, retirement planning, and understanding the benefits and resources available to them.

Supporting veterans interested in entrepreneurship is also crucial. Offering targeted resources and mentorship programs can encourage their business ventures. This can involve providing access to capital, business planning assistance, mentorship networks, and guidance on navigating government contracting opportunities.

Fostering employer engagement is essential. Encouraging employers to recognize the value of hiring veterans and providing incentives for veteran recruitment and retention can enhance employment opportunities. This can include offering tax credits, promoting veteran-friendly workplace practices, and fostering a supportive work environment that values veterans' unique skills and experiences.

By implementing these strategies, stakeholders can contribute to a smoother transition for veterans into civilian employment, providing them with the necessary support, resources, and opportunities to thrive in their chosen careers.

4. Mental and Emotional Well-Being: Financial stability positively impacts veterans' mental and emotional well-being during their transition. Financial stress and instability can contribute to feelings of anxiety, depression, and hopelessness, affecting their overall mental health. On the other hand, financial stability promotes a sense of security, control, and confidence, reducing stress levels and improving mental well-being. With financial

> stability, veterans can focus on their personal growth, pursue hobbies, engage in social activities, and maintain a healthy work-life balance. Financial stability provides the freedom to enjoy life without constant worry about financial constraints, thus enhancing veterans' overall well-being.

Financial stability plays a crucial role in the overall well-being of veterans, including their mental and emotional health. Veterans often face unique challenges when transitioning from military service to civilian life, and financial stability can significantly influence their mental and emotional well-being. Financial stability can alleviate stress and anxiety associated with financial insecurity. Having the necessary financial resources to meet basic needs, cover expenses, and plan for the future reduces the burden of financial worries, leading to improved mental well-being.

Having financial stability empowers veterans with a sense of control over their lives, enabling them to make choices that align with their values and aspirations. This sense of control enhances feelings of self-worth, autonomy, and confidence, positively impacting their mental well-being.

Promoting financial stability contributes to mitigating mental health disparities among veterans. Adequate financial resources ensure access to mental health services, therapies, and treatments, bridging the gap in health-care accessibility and promoting mental well-being.

Financial stability can strengthen social connections and support networks. With reduced financial stress, veterans may feel more comfortable engaging in social activities, maintaining relationships, and seeking emotional support—all of which contribute to improved mental well-being.

Experiencing financial instability can result in chronic stress and anxiety, which can have adverse effects on the mental well-being of veterans. Worries about meeting financial obligations, debt, or inadequate resources for basic needs can contribute to increased psychological distress, creating a constant source of mental strain.

Lack of financial stability can restrict veterans' access to essential health-care services, including crucial mental health support. Without adequate financial resources, veterans may face challenges in affording insurance coverage, co-pays, or specialized mental health treatments. This limited access to health care further exacerbates mental health issues and hampers their ability to seek the support they need.

The presence of financial instability can exert pressure on family dynamics, leading to conflicts within the household. The inability to provide for family needs or fulfill financial responsibilities can lead to emotional distress for veterans and their loved ones. The resulting negative home environment can further impact veterans' mental well-being, creating additional stressors and emotional strain.

Financial instability may curtail veterans' opportunities for self-care activities that promote mental and emotional well-being. Limited financial resources can restrict their ability to engage in hobbies, recreational activities, or therapies that contribute to stress reduction and overall well-being. This lack of access to self-care can further exacerbate the negative impacts on veterans' mental health.

Having financial stability guarantees that veterans can obtain essential mental health services without facing financial obstacles. Providing financial support for mental health treatments, counseling, and therapies can significantly improve veterans' mental well-being and help prevent the development of more severe mental health conditions.

Equipping veterans with financial education and budgeting skills is important. Teaching strategies for saving, debt management, and budgeting helps veterans maintain financial stability, reducing stress and promoting mental well-being. By empowering veterans with the knowledge and tools to manage their finances effectively, they can better navigate the challenges and uncertainties that may arise.

Facilitating employment opportunities and career advancement programs for veterans promotes financial stability and, consequently, mental and emotional well-being. Assisting veterans in finding stable employment, developing job skills, and accessing career resources enhances their financial security and overall life satisfaction. Stable employment not only provides financial stability but also fosters a sense of purpose, accomplishment, and fulfillment.

Implementing holistic support programs that address both the financial and mental health needs of veterans is crucial. By combining financial counseling, mental health support, and social services, these programs can provide comprehensive support to veterans, promoting their overall well-being and helping them thrive in their civilian lives.

By recognizing the importance of financial stability and implementing appropriate support measures, society can contribute to the mental well-being of veterans and ensure they receive the assistance they need to lead fulfilling and successful lives.

5. Family Support and Stability: Financial stability is vital for veterans in supporting their families and maintaining family stability. During the transition, veterans may face the responsibility of providing for their spouse, children, or other dependents. Financial stability ensures the ability to meet family needs, including education, health care, and daily expenses, thereby strengthening family relationships and stability. By providing for their families, veterans can establish a sense of purpose, belonging, and fulfillment, contributing to their overall successful transition and integration into civilian life.

Financial stability enables veterans to meet their family's basic needs, including essential requirements like food, housing, health care, and education. Having access to these necessities fosters a sense of security and stability within the family unit, reducing stress and anxiety. When families are free from the burden of financial insecurity, they can focus on their well-being and enjoy a higher quality of life.

Financial stability has a positive impact on family relationships. Reducing financial stressors and associated conflicts, allows veterans to be more emotionally available and supportive within their families. This promotes stronger bonds, healthier communication, and a more supportive environment for everyone involved. Financial stability enables veterans to create a nurturing atmosphere where family members can thrive and support one another.

Having financial stability allows veterans to pursue opportunities for investing in their family's future. Whether it's planning for their children's education, saving for emergencies, or participating in meaningful family activities, having the necessary financial resources provides families with a sense of unity and shared experiences. These opportunities contribute to the overall well-being of the family and create lasting memories that strengthen their bond.

Ensuring financial stability is vital for veterans and their families, as it grants them access to critical health-care resources and insurance coverage. With adequate financial resources, veterans and their families can afford medical services, medications, mental health support, and preventive care. This ensures the well-being and stability of the entire family, allowing them to maintain good health and address any health-care needs that may arise.

Financial stability enables veterans to support their children's educational pursuits and invest in their own ongoing education and skill development. By having the necessary financial resources, veterans can provide their children with access to quality education and training opportunities, improving their prospects and enhancing the overall financial stability of the family. Education plays a vital role in opening doors to better opportunities for both the veterans and their children, leading to increased financial security and overall family well-being.

Additionally, financial stability plays a crucial role in securing stable housing for veterans and their families. With adequate income, veterans can afford safe and suitable housing, providing a stable and nurturing environment for their children. Stable housing contributes to overall family well-being by ensuring a sense of security, promoting positive family dynamics, and offering a place to call home.

Financial stability alleviates financial stress within the family unit, allowing veterans and their loved ones to focus on other aspects of their lives. Reduced financial stress contributes to improved mental health, fosters healthier relationships, and enhances overall family well-being. Without the burden of constant financial worries, families can allocate their energy toward nurturing positive communication, building emotional connections, and creating a supportive and harmonious family environment.

Having financial stability creates a sense of security that can improve the emotional well-being of veterans and their families. Knowing that their financial needs are met and that they have the resources to cover expenses and pursue their goals instills a sense of confidence and stability.

This emotional security allows families to prioritize their emotional well-being, strengthening their bond, and promoting a positive and nurturing family dynamic.

With financial stability, families can strategically prepare for the future, encompassing retirement, emergencies, and long-term financial objectives. Having the financial resources and stability to prepare for the future instills a sense of preparedness and confidence. Families can make informed decisions, set realistic financial goals, and work toward achieving them, which contributes to a positive family outlook and overall quality of life.

Providing financial counseling and education programs tailored to veterans and their families is crucial in equipping them with the skills and knowledge necessary to effectively manage their finances. By offering this support, veterans and their families can make informed decisions, develop sound financial strategies, and work toward long-term stability. These programs empower them to take control of their financial situation, improve financial literacy, and make wise financial choices, leading to overall improved financial well-being.

Facilitating employment opportunities, job training, and career development programs for veterans play a significant role in enhancing their financial stability and promoting family support and stability. Access to stable employment with fair wages and opportunities for career advancement not only provides financial security, but also positively impacts the entire family. It allows them to meet their financial obligations, plan for the future, and create a stable and supportive environment for their loved ones.

In addition to individual support, establishing social-support networks and community resources specifically targeted toward veterans and their families is essential. These networks can offer financial assistance programs, counseling services, and connections to other essential resources. By creating a sense of belonging and promoting mutual support, these networks provide veterans and their families with access to social support, guidance, and additional financial resources when needed, contributing to their overall well-being and financial stability.

Financial stability is a critical component of successful transition and reintegration for veterans. It enables them to meet their basic needs, pursue education and training, secure employment, enhance their mental well-being, and provide for their families. Financial stability during this crucial period empowers veterans to make informed choices, focus on personal growth and development, and navigate the challenges of transitioning into civilian life. By recognizing and supporting the importance of financial stability for veterans, society can contribute to their successful transition, improve their overall well-being, and honor their invaluable service to the nation.

Veterans' financial responsibility plays a significant role in their ability to afford tuition fees for education and training programs. Limited financial resources may impose restrictions on the institutions or programs they can access, affecting their educational choices and opportunities. Making responsible financial decisions and managing their finances effectively becomes crucial for veterans to navigate the financial challenges associated with pursuing education.

The availability of scholarships and grants is heavily influenced by veterans' financial responsibility. Responsible financial planning and management increase their chances of qualifying for financial assistance, making education more affordable and accessible. By demonstrating financial responsibility, veterans can enhance their eligibility for scholarships, grants, and other forms of financial aid, opening doors to educational opportunities they may not have otherwise been able to pursue.

Financial responsibility also encompasses considering the costs associated with transitioning from military service to educational pursuits. Veterans must account for expenses such as application fees, test preparation materials, and potential relocation costs when making decisions about pursuing education and training opportunities. Being financially responsible requires veterans to carefully assess and plan for these transition costs, ensuring they can navigate the financial challenges that arise during this period.

Veterans' financial responsibility significantly impacts their educational journey, affecting their access to institutions, eligibility for financial aid, and ability to manage the costs of transitioning from military service. By practicing responsible financial planning and management, veterans can overcome financial barriers and create more opportunities for themselves in their pursuit of education and training.

Financial responsibility extends beyond tuition fees to encompass the affordability of housing and utilities while pursuing education or training. Rent, mortgage payments, and utility bills are factors that contribute to veterans' overall financial responsibility and may influence their choice of educational programs or institutions. Being financially responsible requires veterans to carefully consider the affordability of housing and utilities, ensuring that they can meet these obligations while pursuing their educational goals.

Another aspect of financial responsibility for veterans is the cost of books and supplies. Veterans are responsible for covering the expenses related to textbooks, supplies, and other materials necessary for their education and training. These costs can accumulate significantly, impacting their overall financial responsibility and potentially affecting their ability to fully engage in their chosen programs. Managing these expenses and finding cost-effective solutions for obtaining books and supplies becomes essential for veterans in their pursuit of education.

Transportation and commuting costs also fall within the realm of financial responsibility. Veterans must consider the expenses associated with transportation to and from educational or training institutions. Whether it involves budgeting for fuel, public transportation fares, or vehicle maintenance costs, veterans need to allocate funds for these transportation expenses. These costs can influence their decision-making process when selecting educational programs or determining the feasibility of commuting to their chosen institutions.

Financial responsibility encompasses various aspects related to the affordability of housing and utilities, the costs of books and supplies, and transportation expenses while pursuing education or training. Veterans must be mindful of these financial considerations to ensure they can meet their financial obligations and make informed decisions that support their educational goals.

Exercising financial responsibility necessitates veterans to approach their educational investments with thoughtful deliberation and informed decision-making. They must assess the potential return on investment, evaluate the costs and benefits of different programs, and take into account the long-term financial implications of their educational choices. Being financially responsible means being proactive in making choices that align with their goals and financial capabilities.

One aspect of financial responsibility for veterans is managing student loans and debt. Veterans need to take responsibility for understanding the terms of their loans, making informed borrowing decisions, and planning for repayment. Responsible debt management is crucial to avoid financial hardships and maintain financial stability in the long run. By being proactive in understanding their loan obligations and implementing effective repayment strategies, veterans can minimize the impact of student loans on their financial well-being.

Balancing work and education is another important aspect of financial responsibility. Veterans often find themselves in a position where they need to juggle employment and educational pursuits. They may have to make decisions regarding employment opportunities, working hours, and income generation to support their educational goals. Finding the right balance between work and education is essential to ensure they can meet their financial responsibilities while progressing in their educational endeavors.

Accessing financial-aid programs and scholarships is crucial for veterans in managing their financial responsibilities while pursuing education and training. These resources serve as a valuable means to reduce the financial burden, making education more affordable and accessible. By exploring and utilizing financial aid options, veterans can alleviate the financial responsibilities associated with their educational pursuits.

To further support veterans in managing their financial responsibilities, it is important to provide them with tailored financial education and counseling services. These programs can equip veterans with the necessary knowledge and skills to effectively handle their financial obligations. By receiving education on topics such as budgeting, debt management, and long-term financial planning, veterans can make informed decisions regarding their education and training while maintaining financial stability.

Some employers offer assistance programs specifically designed to support veterans in their educational pursuits. These programs may include tuition reimbursement, educational assistance programs, or scholarships. By taking advantage of these opportunities, veterans can reduce their financial responsibilities and enhance their educational opportunities.

Accessing financial-aid programs and scholarships, providing tailored financial education and counseling services, and taking advantage of employer assistance programs are important strategies to help veterans manage their financial responsibilities while pursuing education and training. By leveraging these resources and opportunities, veterans can navigate their educational journeys with greater financial ease and achieve their educational goals.

Veterans' financial responsibility is intricately connected to their educational pursuits and training. By understanding the financial costs, making informed decisions, accessing support and resources, and managing their financial responsibilities effectively, veterans can enhance their educational experiences, expand their career opportunities, and support their families during a critical time of transition.

Chapter 7

Education and Training

The transition from military service to civilian life is a significant milestone for veterans, marked by numerous challenges and opportunities. Among the various factors that contribute to a successful transition, education and training play a pivotal role. Equipping veterans with the necessary skills, knowledge, and credentials not only empowers them to secure meaningful employment, but also fosters personal growth, smooth integration into civilian society, and overall well-being.

1. Enhancing Employability: One of the primary advantages of education and training for veterans is the substantial boost it provides to their employability. Many veterans possess a unique set of skills acquired during their military service, but these skills may not always align directly with civilian job requirements. By pursuing further education and training, veterans can bridge this skills gap, making themselves more competitive in the job market. For instance, acquiring a degree or professional certification can provide veterans with the necessary credentials to pursue careers in fields such as health care, engineering, IT, or business administration.

One of the most significant implications of education for veterans is the enhancement of qualifications and skills. While military service equips veterans with valuable skills such as leadership, teamwork, and problem-solving, these skills may need to be complemented with specialized knowledge and certifications to meet the demands of civilian employment. Pursuing education, whether through degree programs, vocational training, or professional certifications, allows veterans to acquire new skills and upgrade existing ones, making them more competitive candidates in the job market. By aligning their educational pursuits with industry trends and employer requirements, veterans can significantly broaden their employment prospects.

As veterans transition from military service to civilian life, enhancing their qualification skills becomes a crucial step in securing employment and building successful careers. The unique skills and experiences gained during military service provide a strong foundation, but additional qualifications are often necessary to meet the demands of the civilian job market.

One of the primary benefits of enhancing qualification skills for veterans is the ability to fill the skills gap that may exist between military and civilian employment. While veterans possess valuable attributes like leadership, discipline, and teamwork, some technical skills acquired during military service may not directly align with civilian job requirements. By pursuing further education, veterans can acquire new skills or upgrade existing ones, ensuring they possess the competencies demanded by prospective employers. Whether it is learning new technologies, industry-specific processes, or specialized knowledge, enhancing qualification skills enables veterans to bridge the gap and become more competitive in the job market.

Military training equips veterans with a diverse range of skills that can be successfully transferred to civilian contexts. However, it is essential to translate these skills effectively to make them relevant and understandable to civilian employers. Enhancing qualification skills involves articulating military experiences in a way that resonates with civilian employers, highlighting their value and applicability to specific roles. For example, veterans with logistics experience can pursue certification programs in supply chain management to demonstrate their expertise in a civilian context. By enhancing their qualifications, veterans can effectively communicate their transferable skills and increase their chances of securing employment in industries aligned with their military background.

The job market is dynamic, with industry demands evolving rapidly. To remain competitive, veterans must stay abreast of these changes and adapt their qualification skills accordingly. Continuing education and professional development opportunities are key to ensuring that veterans remain current and relevant in their chosen fields. By participating in industry-specific training, attending workshops, or pursuing advanced degrees, veterans can expand their knowledge base and acquire the latest skills demanded by employers. This adaptability not only enhances their employability, but also positions them for growth and advancement within their respective industries.

Enhancing qualification skills opens doors to new opportunities for veterans, providing them with access to a wider range of career options. By acquiring additional qualifications, veterans can explore industries and roles that align with their interests and goals, beyond their military experience. For instance, a veteran with a background in communications can enhance their qualifications by pursuing a degree in public relations or journalism, unlocking opportunities in media, marketing or public relations. By diversifying their skill set and qualifications, veterans can tap into new industries and pursue careers that offer long-term satisfaction and growth.

Enhancing qualification skills is a pivotal step in the successful transition of veterans from military service to civilian employment. By filling the skills gap, transferring military skills to civilian contexts, adapting to changing industry demands, and unlocking new opportunities, veterans can position themselves for long-term success and personal fulfillment. Recognizing the significance of targeted education, training, and certification programs, society must support and invest in accessible resources, financial assistance, and tailored opportunities that empower veterans to enhance their qualification skills. By doing so, we can ensure that veterans have the tools and opportunities they need to thrive in their postservice careers and contribute to the growth and prosperity of their communities.

Transitioning from military service to the civilian workforce can present veterans with a skills gap that may hinder their employment opportunities. Many military occupations have unique technical skills that may not directly translate to civilian jobs. Education serves as a bridge to fill this gap by providing veterans with the opportunity to acquire the skills and knowledge that are in demand within their chosen industries. For example, veterans with experience in logistics or engineering can pursue educational programs in supply chain management or civil engineering, respectively, to enhance their employability in related civilian roles. Education equips veterans with the industry-specific skills that employers seek, improving their chances of securing meaningful employment.

One of the key challenges veterans face when entering the civilian workforce is translating their military skills into terms that are easily understandable and relevant to civilian employers. Education and training programs provide a platform for veterans to bridge this communication gap effectively. These programs offer guidance and resources to help veterans articulate their military experiences and skills in a way that resonates with civilian employers. By assisting in resume writing, interview preparation, and professional development, education and training initiatives help veterans highlight the transferable skills acquired during their military service, making them more marketable to potential employers.

Military training often provides veterans with a strong foundation in discipline, leadership, and teamwork. However, to excel in civilian careers, veterans may require industry-specific knowledge and expertise. Education and training programs offer targeted courses and certifications that enable veterans to acquire the industry-specific skills demanded by employers. Whether it is pursuing a degree in a specific field, attending vocational training programs, or obtaining specialized certifications, veterans can enhance their knowledge and capabilities, bridging the skill gap between military service and civilian employment. These programs ensure that veterans are equipped with the latest industry practices, technology, and trends, making them valuable assets in their chosen fields.

Technological advancements and evolving industry practices mean that certain technical skills acquired during military service may become outdated over time. Education and training initiatives for veterans address this challenge by providing opportunities for veterans to update and enhance their technical skills. Whether it involves learning new software applications, mastering advanced machinery, or understanding cutting-edge techniques, education and training programs keep veterans abreast of the latest advancements in their respective industries. By closing the technical-skill gap, veterans remain competitive and adaptable, ready to meet the demands of the rapidly changing job market.

Many veterans seek to explore new career paths after leaving the military, and education and training programs offer a pathway for a successful transition. These programs provide veterans with the opportunity to acquire skills and knowledge in fields unrelated to their military background, opening up a range of new career possibilities. For instance, a veteran with combat medical experience may pursue a nursing degree to transition into the health-care industry. By offering the necessary education and training, these programs enable veterans to bridge the skill gap and successfully transition into civilian careers in industries that align with their interests and aspirations.

Education and training initiatives designed for veterans play a pivotal role in bridging the skill gap that exists between military service and civilian employment. By assisting veterans in translating their military skills, acquiring industry-specific knowledge, updating technical competencies, and facilitating transitions to new career paths, these programs empower veterans to bridge the gap effectively. Recognizing the significance of tailored education and training for veterans, it is essential to continue supporting and investing in accessible resources, scholarships, and career counseling to ensure that veterans have the tools and opportunities they need to succeed in their postservice careers. By doing so, we can bridge the skill gap, unlock veterans' potential, and harness their valuable contributions to the civilian workforce.

2. Expanding Career Options: Education opens up a world of new career possibilities for veterans. It allows them to explore different industries and sectors beyond their military experience, broadening their career options. Veterans can leverage their transferable skills in conjunction with education to pursue careers that align with their interests, passion, and long-term goals. Whether it is transitioning to a civilian role related to

their military specialization or embarking on a completely new career path, education equips veterans with the knowledge and credentials to explore diverse employment opportunities. By expanding their horizons and venturing into new fields, veterans increase their chances of finding fulfilling and sustainable employment postservice.

As veterans transition from military service to civilian life, one of the significant challenges they face is expanding their career options beyond their military background. Veterans' education plays a crucial role in this process by equipping them with the necessary knowledge, skills, and qualifications to explore diverse career paths in the civilian market.

Veterans possess a wide range of transferable skills acquired during their military service, such as leadership, teamwork, problem-solving, and adaptability. Veterans' education enables them to leverage these skills and apply them to different industries and roles in the civilian market. By pursuing education, veterans can enhance their transferable skills, making them attractive candidates for a broader range of career options. For instance, a veteran with leadership experience in the military can further develop their management skills through education, enabling them to pursue leadership roles in various sectors such as business, health care, or nonprofit organizations. Education empowers veterans to showcase their transferable skills and opens doors to new and exciting career possibilities.

One of the key steps in leveraging transferable skills is effectively translating military experiences into the language and requirements of the civilian job market. Veterans should identify the skills they developed during their military service and articulate them in a way that resonates with civilian employers. For example, a veteran's leadership experience in the military can be presented as effective project management, decision-making, and team-coordination skills. By understanding how to bridge the gap between military and civilian terminology, veterans can effectively communicate their transferable skills to potential employers, increasing their chances of securing employment in diverse industries.

Soft skills, such as communication, teamwork, adaptability, and problem-solving, are highly sought after by employers in today's job market. Veterans often excel in these areas due to their training and experiences in the military. By highlighting these soft skills on resumes, cover letters, and during interviews, veterans can showcase their ability to work well in diverse teams, handle pressure, and solve complex problems. Employers recognize the value of such skills and appreciate the unique perspective and work ethic that veterans bring to the table. By effectively emphasizing their soft skills, veterans can position themselves as strong candidates for a wide range of job opportunities.

Leadership skills acquired during military service are among the most valuable transferable skills veterans possess. Whether it is leading a small team or managing complex operations, veterans are accustomed to making critical decisions, motivating others, and accomplishing objectives. These leadership skills can be leveraged in civilian careers, where veterans can assume leadership roles or contribute as effective team members. By providing specific examples of leadership experiences and accomplishments, veterans can demonstrate their ability to lead, mentor, and inspire others, thereby enhancing their appeal to potential employers seeking individuals with strong leadership qualities.

The military environment demands adaptability and problem-solving skills, as service members often face unpredictable and challenging situations. Veterans can leverage these skills in the civilian market by highlighting their ability to think critically, make sound judgments, and adapt quickly to changing circumstances. The ability to navigate complex problems and find effective solutions is highly valuable in various industries, and veterans can demonstrate their proficiency in these areas through real-life examples from their military experiences. Employers appreciate candidates who can handle ambiguity and find innovative solutions, making veterans with transferable skills highly desirable in today's dynamic work environments.

Veterans possess a diverse range of transferable skills that can be leveraged to excel in civilian careers. By effectively translating military skills to civilian contexts, highlighting soft skills, demonstrating leadership abilities, and emphasizing adaptability and problem-solving skills, veterans can position themselves as valuable assets to employers across various industries. Veterans need to recognize the value of their transferable skills and present them in a way that resonates with civilian employers. Additionally, employers and society at large should appreciate and acknowledge the unique abilities and experiences that veterans bring to the workforce. By leveraging their transferable skills, veterans can unlock their full potential and contribute significantly to the civilian job market.

While veterans bring valuable skills to the civilian job market, certain industries may require specific knowledge or qualifications that veterans may not possess. Veterans' education addresses this gap by providing opportunities to acquire industry-specific knowledge and expertise. By pursuing degrees, certifications, or vocational training programs, veterans can gain specialized knowledge in fields that interest them. This expansion of knowledge allows veterans to explore new career options and compete effectively in industries that align with their passions and long-term goals. Whether it is transitioning to a technical field, business sector, or creative industry, veterans' education broadens their horizons and enhances their employability in diverse career paths.

In addition to traditional academic pathways, veterans can acquire industry-specific knowledge through vocational training programs and certifications. These programs are designed to provide practical, hands-on training in specific industries or occupations. Veterans can opt for vocational training programs that focus on fields such as construction, information technology, automotive, health care, and more. These programs typically offer specialized courses and workshops that equip veterans with the skills and certifications required for entry-level positions in their target industries. By completing vocational training and earning industry-specific certifications, veterans demonstrate their commitment to continuous learning and acquire the necessary credentials to establish themselves as competent professionals in specific fields.

Vocational training and certifications play a vital role in empowering veterans as they transition from military service to civilian careers. These programs provide veterans with the necessary skills, knowledge, and industry-specific qualifications to thrive in various occupations and industries.

Vocational training programs offer practical, hands-on learning experiences that directly align with specific occupations and industries. Veterans can benefit from the focused and targeted nature of vocational training, allowing them to quickly acquire the skills required for entry-level positions. Whether it's in fields such as construction, automotive technology, health care, information technology, or culinary arts, veterans can gain industry-relevant skills through vocational training that can be immediately applied in the workplace. The emphasis on practical skills development enables veterans to demonstrate their competence and readiness to contribute to their chosen industries from day one.

Veterans benefit from vocational training programs as they gain industry-specific knowledge and qualifications that are crucial for pursuing careers in specialized fields. These programs offer specialized coursework and training that equip veterans with the technical expertise and understanding of industry practices. By completing vocational training, veterans can earn certifications that validate their skills and knowledge, making them more attractive to employers seeking candidates with specialized training. These certifications serve as tangible evidence of veterans' competence and commitment to their chosen fields, enhancing their employability and opening doors to a wide range of job opportunities.

Transitioning from military service to civilian careers can often involve bridging skill gaps that exist between the two domains. Vocational training programs play a crucial role in addressing these gaps by equipping veterans with the specific skills and knowledge required in civilian occupations. For instance, veterans with mechanical or technical skills acquired in the military can enroll in vocational training programs focused on civilian trades, such as plumbing, electrical work, or HVAC systems. Through these programs, veterans can gain the necessary civilian certifications and skills to seamlessly transition into these occupations, addressing skill gaps and ensuring a smooth career transition.

Vocational training programs often provide tailored support and career guidance to veterans, recognizing their unique needs and experiences. These programs understand the challenges veterans may face during their transition and provide resources such as career counseling, job- placement assistance, and mentorship opportunities. The personalized support ensures that veterans receive guidance on selecting the most appropriate vocational training program based on their skills, interests, and career goals. Additionally, career services offered by vocational training institutions help veterans navigate the job market, refine their resumes, and prepare for interviews, increasing their chances of securing employment in their chosen fields.

Compared to traditional college degrees, vocational training programs often present veterans with cost-effective and time-efficient alternatives. These programs are typically shorter in duration, allowing veterans to quickly acquire the necessary skills and qualifications without spending several years in education. Additionally, many vocational training programs offer financial assistance, grants, scholarships, or tuition reimbursement opportunities specifically tailored to support veterans. These benefits make vocational training a viable and accessible option for veterans seeking to enter the civilian workforce promptly and efficiently.

Vocational training and certifications provide invaluable opportunities for veterans to acquire industry-specific skills, knowledge, and qualifications. Through practical-skills development, industry-specific training, bridging skill gaps, tailored support, and cost-effective options, veterans can enhance their employability and succeed in civilian careers. Veterans must explore vocational training programs that align with their interests, leverage the resources available, and tap into the supportive networks within these programs.

Mentorship and networking play a significant role in acquiring industry-specific knowledge. Veterans can seek out mentors who are experienced professionals in their desired fields. A mentor can provide guidance, share insights, and offer valuable industry-specific knowledge and advice. Networking events, industry conferences, and professional associations provide opportunities for veterans to connect with professionals who have expertise in their target industries. By actively networking and building relationships, veterans can tap into the knowledge and experiences of industry insiders, gaining valuable insights into industry-specific practices, trends, and opportunities.

Mentorship and networking provide a practical and personalized approach to acquiring industry-specific knowledge, enabling veterans to navigate their chosen fields more effectively.

Guiding the Path to Success: Mentorship offers veterans the opportunity to learn from experienced professionals who can provide invaluable insights, advice, and guidance. Mentors act as trusted advisors, sharing their knowledge, experiences, and expertise to help veterans navigate the challenges and complexities of their chosen industries. Through mentorship, veterans gain access to a wealth of practical knowledge; learn industry-specific best practices; and receive personalized career advice tailored to their goals. Mentors can help veterans identify potential career paths, expand their professional networks, and offer guidance on skill development and advancement opportunities. The guidance provided by mentors can significantly enhance veterans' confidence, accelerate their learning curve, and contribute to their long-term career success.

Expanding Opportunities and Connections: Networking is a powerful tool that allows veterans to expand their professional connections and access a wider range of opportunities. By actively participating in networking events, industry conferences, and professional associations, veterans can engage with professionals from various sectors, build relationships, and establish a strong professional network. Networking provides veterans with exposure to diverse perspectives, industry trends, and potential job openings. Through networking, veterans can gain insights into specific industries, learn about emerging career opportunities, and connect with potential employers or mentors. Building a robust professional network creates a platform for collaboration, knowledge sharing, and ongoing career growth.

Veteran-Specific Networking Opportunities: Recognizing the unique experiences and challenges faced by veterans, numerous organizations and initiatives provide veteran-specific networking opportunities. Veteran-focused professional associations, online platforms, and networking events facilitate connections among veterans and employers, creating a supportive ecosystem for career development. These platforms and events bring together veterans, industry professionals, recruiters, and mentors who understand the value veterans bring to the civilian workforce. By participating in veteran-specific networking opportunities, veterans can connect with individuals who appreciate their military background, share common experiences, and provide relevant guidance tailored to their transition into civilian careers.

Mentoring Others and Paying It Forward: Veterans possess a wealth of knowledge, leadership skills, and unique experiences that make them excellent mentors for fellow veterans and aspiring professionals. By engaging in mentorship, veterans not only receive guidance but also have the opportunity to give back to their community and support others in their career journeys. Sharing their experiences, insights, and lessons learned allows veterans to make a meaningful impact on the lives of mentees and contribute to the professional development of the next generation. Through mentorship, veterans can help shape the future of their industries by fostering a culture of continuous learning, resilience, and excellence.

Mentorship and networking are invaluable resources for veterans transitioning into civilian careers. By connecting with mentors, veterans can gain guidance, industry-specific knowledge, and personalized career advice, accelerating their professional growth. Networking opens doors to new opportunities, expands professional connections, and provides exposure to industry trends and job openings. Veteran-specific networking initiatives offer tailored support and a sense of community during the transition process. As veterans engage in mentorship and networking, they not only receive support but also have the opportunity to pay it forward by mentoring others, contributing to the collective success and empowerment of the veteran community.

Acquiring industry-specific knowledge is essential for veterans as they transition into civilian careers. By pursuing higher education, vocational training, and certifications; engaging in professional development activities; and leveraging mentorship and networking opportunities, veterans can acquire the industry-specific knowledge and qualifications necessary to thrive in their chosen fields. Veterans need to recognize the value of continuous learning and actively seek opportunities to enhance their industry-specific knowledge. Additionally, employers and society should provide support and resources that facilitate veterans' access to industry-specific training and education, recognizing the valuable skills and experiences veterans bring to the civilian job market.

Education programs provide veterans with valuable opportunities to network and connect with professionals in various industries. These connections can be instrumental in expanding career options as they offer insights, mentorship, and potential job opportunities. Through educational institutions, veterans can attend career fairs, industry events, and networking sessions, allowing them to build relationships with employers, industry experts, and fellow students. These networks can help veterans explore different career options, gain insider knowledge about specific industries, and access hidden job markets. The connections forged during veterans' education can significantly expand their career options by opening doors to unexplored possibilities and providing valuable support and guidance during their transition into civilian employment.

Professional networks play a crucial role in career development, job search, and business opportunities. For veterans transitioning to civilian careers, building and nurturing such networks becomes paramount. These networks provide avenues for mentorship, access to job opportunities, and professional growth. By connecting with fellow professionals, veterans can gain valuable insights into industry trends, best practices, and potential career pathways.

One of the unique advantages veterans have is their extensive military network. Veterans can tap into this network to connect with individuals who share a common background and understanding of their experiences. Military networks can provide a supportive community that offers guidance, advice, and assistance in navigating the civilian job market. By utilizing these connections, veterans can widen their professional circles and open doors to new opportunities.

Numerous organizations and initiatives are dedicated to supporting and empowering veterans in their professional pursuits. These organizations offer resources, mentoring programs, and networking events specifically tailored to veterans. For instance, organizations like Hiring Our Heroes, American Corporate Partners, and the Veteran Jobs Mission provide veterans with access to networking opportunities, career counseling, and job fairs. By actively participating in these organizations, veterans can engage with industry professionals, build relationships, and expand their professional networks.

Professional networks provide veterans with access to mentors who can guide them through the intricacies of the civilian professional world. Mentors, who may be fellow veterans or industry professionals, offer valuable advice, share their experiences, and provide career guidance. Mentoring relationships established through professional networks can help veterans develop new skills, broaden their perspectives, and make informed decisions about their career paths.

Veterans bring a wealth of experience, skills, and attributes to the professional world, making them valuable assets in various industries. However, a successful transition to civilian careers often relies on the development of professional networks and connections. By actively fostering these networks, veterans can tap into valuable resources, access job opportunities, and receive guidance from mentors. Simultaneously, the broader society benefits from the unique perspectives and contributions veterans bring to diverse professional environments. Recognizing the importance of veterans' professional networks and providing support in building and maintaining them will lead to increased career success and opportunities for our veterans.

Education also nurtures the entrepreneurial spirit within veterans, providing them with the knowledge and skills to start their businesses or pursue self-employment opportunities. Veterans' education offers courses and programs focused on entrepreneurship, business management, and innovation, equipping veterans with the necessary tools to navigate the world of entrepreneurship. By acquiring business acumen, marketing strategies, and financial literacy, veterans can explore entrepreneurial endeavors in various sectors, leveraging their unique skill sets and experiences. Education empowers veterans to chart their career paths, create employment opportunities for themselves and others, and expand their career options beyond traditional employment models.

The skills and traits acquired during military service are highly transferable to the entrepreneurial realm. Veterans are accustomed to working in high-pressure environments, making quick decisions, and adapting to changing circumstances. Their ability to lead teams, communicate effectively, and solve complex problems are invaluable assets in the world of entrepreneurship. Veterans often possess a strong work ethic, discipline, and perseverance, which are all crucial qualities for building and growing a business.

Veterans have firsthand experience in overcoming challenges and adversity, which instills in them a sense of resourcefulness and resilience. These qualities are essential for entrepreneurs who often face setbacks, uncertainty, and obstacles along their journey. Veterans' ability to adapt to new situations, improvise, and find innovative solutions can give them a competitive edge in the entrepreneurial landscape.

Numerous programs and initiatives have been established to support veteran entrepreneurs. Organizations like the US Small Business Administration's Office of Veterans Business Development (OVBD), Veterans Business Outreach Centers (VBOCs), and the Veterans Entrepreneurship Program (VEP) offer resources, training, mentoring, and financial assistance specifically tailored to veterans interested in starting their businesses. These programs provide veterans with the necessary guidance and support to navigate the entrepreneurial world successfully.

Veterans have a vast network of fellow service members, military personnel, and veterans who can be instrumental in their entrepreneurial endeavors. These connections can provide valuable advice, mentorship, and access to resources. Veterans often share a strong camaraderie and are willing to support each other's businesses. Leveraging their military networks can help veterans gain credibility, expand their customer base, and secure partnerships or collaborations.

Entrepreneurship requires discipline, goal-setting, and strategic thinking. Veterans are well-versed in these areas due to their military training. They understand the importance of setting clear objectives, creating actionable plans, and executing tasks with precision. Additionally, veterans are accustomed to following procedures and protocols, which can contribute to building efficient and well-structured businesses.

The military encourages diversity and inclusion, as service members come from various backgrounds, cultures, and experiences. Veterans bring this understanding of diversity and inclusion to their entrepreneurial endeavors, enabling them to create businesses that embrace different perspectives and cater to a broader range of customers. By fostering diverse teams and inclusive business practices, veteran entrepreneurs can tap into new markets and effectively address the needs of diverse populations.

Veterans possess a unique skill set, experience, and mindset that make them well-suited for entrepreneurial endeavors. Their transferable skills, resourcefulness, resilience, access to support programs, and ability to leverage military networks contribute to their success in the business world. By embracing their entrepreneurial spirit, veterans can not only create thriving ventures but also contribute to economic growth, job creation, and innovation. Recognizing and supporting the entrepreneurial aspirations of veterans can lead to a stronger, more dynamic economy while empowering veterans to build fulfilling and impactful careers beyond their military service.

Veterans' education plays a pivotal role in expanding career options for veterans in the civilian market. By leveraging transferable skills, acquiring industry-specific knowledge, fostering professional networks, and nurturing entrepreneurial endeavors, education empowers veterans to pursue diverse and fulfilling career paths. Society must provide accessible and tailored educational resources, financial support, and mentorship to veterans, enabling them to maximize their potential and access a wide range of career opportunities. By recognizing the value of veterans' education in expanding career options, we can honor their service by facilitating their successful transition into civilian life.

Education not only enhances veterans' employability but also positions them for leadership roles within organizations. The leadership skills and experiences gained during military service form a strong foundation for success in managerial and executive positions. However, employers often seek candidates with advanced education to fill these leadership roles. By pursuing higher education, such as a master's degree or an MBA, veterans can demonstrate their commitment to personal growth and development, while also acquiring the knowledge and expertise necessary to excel in leadership positions. Education equips veterans with the strategic thinking, decision-making, and communication skills needed to lead teams and contribute to organizational success.

Military service fosters the development of strong leadership skills. Veterans have undergone extensive training in decision-making, strategic thinking, problem-solving, and effective communication. They have experience leading diverse teams in high-pressure environments, managing resources efficiently, and achieving mission objectives. These skills translate seamlessly into the civilian workforce, making veterans valuable assets in leadership roles.

Veterans are known for their exceptional work ethic and unwavering commitment to achieving goals. In the military, they are accustomed to working long hours, adhering to strict standards, and going above and beyond to accomplish tasks. This level of dedication and commitment is highly valued in leadership positions, as it inspires and motivates teams to perform at their best. Veterans bring a sense of discipline and drive that can have a transformative impact on organizational culture.

Military service instills adaptability and resilience in veterans, enabling them to thrive in dynamic and challenging environments. They have experienced situations that require quick thinking, flexibility, and the ability to overcome obstacles. As leaders, veterans are well-equipped to navigate through uncertainty and change, making them adept at making strategic decisions in complex and rapidly evolving situations. Their resilience also allows them to effectively handle setbacks and inspire others to persevere in the face of adversity.

Veterans understand the importance of teamwork and collaboration in achieving objectives. They have a deep appreciation for the value of each team member's contribution and are skilled at building cohesive and motivated teams. Veterans excel at fostering an environment of trust, respect, and open communication, which enhances collaboration and innovation. Their ability to leverage the strengths of individuals and promote a shared sense of purpose contributes to the overall success of their teams.

Integrity is a core value instilled in military personnel, and veterans bring this commitment to ethical behavior into their leadership roles. They prioritize honesty, accountability, and ethical decision-making, which fosters trust among their teams and stakeholders. Veterans' strong moral compass guides their leadership approach, ensuring they lead with integrity and make principled decisions that align with organizational values.

The military promotes diversity and inclusion, as service members come from various backgrounds, cultures, and experiences. Veterans carry this appreciation for diversity into their leadership roles, valuing different perspectives and fostering inclusive environments. They understand the importance of diverse teams in driving innovation and problem-solving, and they actively advocate for creating opportunities for individuals from underrepresented groups.

Veterans possess a unique combination of skills, experiences, and attributes that make them well-suited for leadership positions. Their leadership skills developed in the military, strong work ethic, adaptability, and commitment to collaboration contribute to their success in advancing to leadership roles. Moreover, their integrity, ethical decision-making, and advocacy for diversity and inclusion bring significant benefits to organizations. Recognizing and embracing the leadership potential of veterans not only strengthens individual organizations but also contributes to a more diverse, inclusive, and effective professional landscape. Veterans bring a wealth of experiences and perspectives that enhance decision-making, innovation, and organizational success, making them invaluable assets in leadership positions.

Education has profound implications for veterans seeking employment opportunities after their military service. By enhancing qualifications and skills, bridging the skills gap, expanding career options, and positioning veterans for leadership roles, education plays a transformative role in empowering veterans to secure meaningful employment and build successful civilian careers. Recognizing the significance of education about employment, it is crucial to provide veterans with accessible and tailored educational resources, financial assistance, and support networks to ensure a seamless transition and a brighter future for those who have dedicated their lives to serving their country.

3. Adapting to the Civilian Workforce: Military service instills a strong work ethic, discipline, and teamwork in veterans,

attributes that are highly valued in the civilian workforce. However, transitioning from the structured and regimented military environment to a more flexible and dynamic civilian workplace can be challenging. Education and training programs specifically tailored for veterans can help them adapt to the civilian work culture, develop effective communication skills, and understand the intricacies of workplace dynamics. This enables veterans to seamlessly integrate into their new work environments, increasing their chances of long-term career success.

Numerous programs and resources have been established to support veterans in their transition to the civilian workforce. For instance, the Department of Defense's Transition Assistance Program (TAP) provides training, counseling, and employment resources to help veterans navigate the job search process. Additionally, organizations like the US Chamber of Commerce Foundation's Hiring Our Heroes and the Department of Veterans Affairs Vocational Rehabilitation and Employment (VR&E) program offer career counseling, job fairs, apprenticeships, and internships to facilitate the transition. By actively utilizing these transitional programs and resources, veterans can access the guidance and support needed to adapt successfully to the civilian workforce.

The civilian workforce often operates with different cultural norms, hierarchies, and communication styles compared to the military. Veterans need to familiarize themselves with the organizational culture, norms, and practices of their new workplace. Adapting to new workplace dynamics, understanding the civilian perspective, and effectively collaborating with colleagues from diverse backgrounds are essential for successful integration into the civilian workforce.

Adapting to the civilian workforce can present challenges for veterans, but with the right strategies and resources, they can successfully navigate this transition. Recognizing and highlighting transferable skills, bridging gaps in qualifications, utilizing transitional programs and resources, seeking mentorship and networking opportunities, embracing lifelong learning, and building cultural awareness are all vital aspects of adapting to the civilian workforce.

4. Opening Doors to New Opportunities: Education and training not only provide veterans with practical skills but also open doors to new opportunities they may not have considered before. These opportunities can range from starting their businesses to pursuing advanced degrees in their areas of interest. By investing in education and training, veterans gain the tools and knowledge necessary to explore different career paths and make informed decisions about their future. Furthermore, education broadens veterans' horizons, exposing them to new ideas, cultures, and perspectives, fostering personal growth and a deeper understanding of the world around them.

Many organizations recognize the value that veterans bring to the workforce and have established initiatives and programs to support their recruitment. Veterans should actively seek out companies and employers that prioritize hiring veterans. They can research organizations that participate in veteran hiring initiatives like the Veteran Jobs Mission, Hiring Our Heroes, or the VOW to Hire Heroes Act. These initiatives create additional opportunities for veterans to connect with employers who value their unique skill sets and experiences.

To support veterans in their career transition, various programs and resources have been established. TAP offers workshops, training, and counseling to help veterans navigate the job search process. Additionally, the VR&E program provides career counseling, job placement assistance, and educational support. These programs offer guidance, resources, and personalized assistance to help veterans identify new employment opportunities that align with their skills and interests.

Many veterans choose to pursue further education or training to enhance their qualifications and expand their career opportunities. The GI Bill provides educational benefits that can be used toward college degrees, vocational programs, and apprenticeships. Veterans can also explore vocational schools, community colleges, and online learning platforms to acquire new skills or certifications. Programs like the Post-9/11 GI Bill, the Yellow Ribbon Program, and the Troops to Teachers Program provide financial assistance and support for veterans pursuing higher education.

Building professional networks and seeking mentorship are invaluable for veterans seeking new employment and educational opportunities. Veterans can connect with fellow service members, veterans, and professionals in their desired field through veteran-focused organizations, online platforms, and networking events. Mentors can provide guidance, advice, and support as veterans navigate the job market or educational pathways. These connections can lead to referrals, insights into job openings, and access to educational resources and scholarships.

Numerous resources are available specifically for veterans seeking employment and education opportunities. Veterans' employment representatives at state employment agencies, career centers at colleges and universities, and veteran service organizations offer guidance, job placement assistance, and access to resources tailored to veterans' unique needs. Veterans should utilize these resources to explore job opportunities, access educational benefits, and receive personalized support throughout their transition.

Veterans possess valuable skills, experiences, and work ethics that make them attractive candidates for both employment and education opportunities. By effectively translating their military experience, leveraging veteran hiring initiatives, utilizing career transition assistance programs, pursuing additional education and training, networking, and accessing veteran-specific resources, veterans can successfully obtain new opportunities for employment and education. The support and resources available to veterans, combined with their determination and resilience, ensure that they can navigate this transition and successfully gain employment.

5. Enhancing Mental Well-Being: The transition from military service to civilian life can be mentally and emotionally challenging for veterans. Education and training programs offer more than just career prospects; they provide a structured and purposeful pursuit that can aid in the transition process

and improve overall mental well-being. Engaging in learning activities helps veterans maintain a sense of purpose, build confidence, and combat feelings of isolation that may arise during the transition period. Additionally, education and training programs often provide access to support networks and resources that can assist veterans in overcoming the various challenges they may face.

Building a strong support network is crucial for maintaining mental health throughout the educational journey. Veterans can connect with fellow student veterans, join campus veterans' organizations, or participate in counseling services specifically tailored for veterans. Engaging with like-minded individuals who understand their unique experiences can provide a sense of camaraderie, support, and encouragement.

Most higher education institutions offer counseling services to students, including veterans. These services can provide support for managing stress, anxiety, depression, or any other mental health challenges. Veterans should take advantage of these resources and seek professional help whenever needed. Campus counselors can offer guidance, coping strategies, and techniques to manage the pressures of academic life effectively.

Balancing academic responsibilities, personal life, and self-care is essential for veterans' mental health. It is important to create a realistic schedule that allows for adequate study time, relaxation, social interactions, and physical exercise. Prioritizing self-care activities, such as meditation, exercise, hobbies, or spending time with loved ones, helps in reducing stress and maintaining a healthy work-life balance.

Higher education can be demanding and stressful, particularly when combined with the challenges of transitioning from military to civilian life. Veterans should develop and practice stress management techniques, such as deep-breathing exercises, mindfulness, or journaling, to help reduce anxiety and promote mental well-being. These techniques can be integrated into daily routines and serve as effective tools for managing academic pressures.

Many institutions provide academic support services, such as tutoring, study groups, or writing centers, to help students succeed in their coursework. Veterans should not hesitate to utilize these resources as they can alleviate academic stress and provide valuable assistance. Seeking help early when encountering challenges or difficulties can prevent the buildup of stress and anxiety.

Maintaining open lines of communication with instructors and peers can contribute to a positive educational experience. If veterans are facing challenges related to their mental health or specific needs, they should feel comfortable discussing these issues with their professors. In many cases, professors can offer accommodations or provide additional support. Similarly, engaging in class discussions and group projects can foster connections with classmates and create a sense of belonging.

Veterans need to allow themselves breaks and practice self-compassion throughout their educational journey. It is natural to feel overwhelmed at times, and taking short breaks to recharge can be beneficial. Additionally, veterans should practice self-compassion by acknowledging their efforts and accomplishments, rather than being overly critical of themselves. Celebrating milestones and recognizing progress can promote a positive mindset and contribute to mental well-being.

While campus resources are vital, veterans should also maintain connections with their existing support networks outside of the academic environment. Family, friends, or fellow veterans can offer additional emotional support and understanding during the educational journey. Regular communication with loved ones can provide a sense of stability and contribute to mental health and well-being.

Education and training serve as essential pillars in the successful transition of veterans from military service to civilian life. By investing in further education, veterans enhance their employability, adapt to the civilian workforce, open doors to new opportunities, and promote their overall well-being. Recognizing the significance of education and training for veterans, society must support and provide accessible resources, scholarships, and tailored programs to ensure a smooth and successful transition for those who have bravely served their country.

Education and training play a critical role in shaping employment opportunities for veterans. The skills, knowledge, and qualifications acquired through education and training programs can significantly enhance veterans' competitiveness in the job market.

Veterans are allowed to acquire new skills and expand their knowledge base through education and training programs. By pursuing higher-education degrees, vocational training, or specialized certifications, veterans can develop expertise in various fields and industries. These enhanced skills and knowledge make veterans more marketable and increase their attractiveness to potential employers. They can demonstrate their ability to adapt and apply their newly acquired skills to meet the evolving demands of the job market.

Many education and training programs have a specific focus on particular industries or occupations. Veterans who pursue education or training aligned with their career goals can gain industry-specific knowledge and skills. This specialization allows veterans to target job opportunities in sectors that match their interests and aspirations. Employers appreciate candidates with relevant expertise, and veterans with specialized training can demonstrate their readiness to contribute effectively in specific roles or industries.

Veterans can benefit from education and training programs as they offer valuable networking opportunities. Veterans attending college or vocational programs have the chance to connect with professors, fellow students, alumni, and industry professionals. These connections can open doors to internships, mentorship programs, and job-placement opportunities. Networking can provide veterans with insights into job openings, industry trends, and career pathways, facilitating their access to employment opportunities.

Education and training programs help veterans further develop their transferable skills, such as leadership, teamwork, problem-solving, and critical thinking. These skills honed through coursework, group projects, and real-world applications are highly valued by employers across various sectors. Education and training programs also offer opportunities for veterans to enhance their communication skills, adaptability, and cultural competency, which are essential for successful integration into the civilian workforce.

Certain careers require specific credentials or certifications. Education and training programs often provide the necessary coursework and preparation for these certifications. By obtaining industry-recognized credentials, veterans can demonstrate their competence and commitment to professional growth. Credentials serve as tangible proof of their qualifications, enhancing their employability and opening doors to employment opportunities that may require specialized knowledge or expertise.

Pursuing education and training programs after military service demonstrates veterans' commitment to personal and professional growth. Employers recognize the dedication and work ethic required to succeed in educational pursuits, and they often view veterans as reliable, disciplined, and driven individuals. Education and training services as evidence of veterans' determination to excel in their chosen fields, positively influencing employers' perceptions and increasing the likelihood of securing employment opportunities.

Education and training programs provide veterans with invaluable opportunities to acquire new skills, expand their knowledge base, and specialize in specific industries. The impact of education and training on veterans' employment opportunities is significant, as they gain relevant expertise, enhance transferable skills, and develop networks within their chosen fields. By pursuing education and training, veterans can maximize their employability, increase their competitiveness in the job market, and successfully transition into the civilian workforce.

Chapter 8

Employment Opportunities

Transitioning from military service to civilian life presents numerous challenges for veterans, and one of the most significant hurdles is finding suitable employment opportunities. Despite possessing a wealth of valuable skills and experiences, veterans often face unique obstacles during their job search. In this chapter, we will explore the struggles veterans encounter when seeking employment during the transition out of the service and discuss potential solutions to address these challenges.

1. Translating Military Experience: One of the primary challenges veterans face is translating their military experience into terms that civilian employers can understand and value. The terminology, acronyms, and job titles used in the military may not align with those commonly used in civilian job descriptions, making it difficult for veterans to effectively showcase their skills and experiences. Additionally, employers may not fully grasp the transferable skills veterans possess, such as leadership, problem-solving, adaptability, and teamwork. Overcoming this hurdle requires veterans to develop the ability to articulate their military experience in civilian terms and demonstrate the relevance of their skills to potential employers.

Veterans bring a wealth of valuable skills and qualities to the civilian job market. Leadership, teamwork, discipline, adaptability, problem-solving, and resilience are just a few examples of the transferable skills honed during military service. Veterans must recognize and articulate the value of their military experience in the context of civilian employment. Understanding how their skills align with the needs of different industries and job roles is the first step in effectively translating military experience.

One of the primary challenges veterans face is bridging the terminology gap between the military and civilian sectors. Military jargon, acronyms, and job titles may be unfamiliar to civilian employers, making it difficult for them to understand the full extent of a veteran's skills and experiences. Veterans must learn to translate their military roles and responsibilities into civilian terms that resonate with employers. This requires identifying the core competencies and achievements from their military service and expressing them in language that hiring managers can easily comprehend and appreciate.

Translating military experience involves showcasing transferable skills that are relevant to civilian employment. Veterans should focus on highlighting skills such as leadership, communication, teamwork, problem-solving, project management, and adaptability. Providing concrete examples of how these skills were utilized in military contexts can help employers understand the applicability of these experiences in a civilian work setting. Veterans should also emphasize their ability to learn quickly, follow procedures, and perform under pressure, as these qualities are highly valued by employers in various industries.

Employers often respond positively to measurable achievements and the impact an individual has made in previous roles. Veterans should quantify their accomplishments whenever possible. This may involve quantifying improvements in efficiency, cost savings, successful mission outcomes, or awards received during military service. By providing quantifiable metrics, veterans can effectively demonstrate their contributions and the value they can bring to potential employers.

To further enhance their translation of military experience, veterans can pursue professional- development opportunities and industry-specific certifications. This demonstrates a commitment to ongoing learning and ensures that veterans stay up-to-date with industry practices and trends. Additionally, certifications help bridge the gap between military and civilian terminology, providing employers with a recognized standard of competence and expertise.

Veterans can leverage various programs and resources designed to assist them in translating military experience for civilian employment. TAPs, veteran-specific job fairs, and organizations specializing in veteran employment support offer invaluable guidance and resources. These programs provide veterans with resume-writing assistance, interview coaching, and networking opportunities with employers who value the unique skills and experiences veterans bring to the table.

2. Lack of Civilian Work Experience: Many veterans transitioning out of the service face the challenge of limited civilian work experience. While their military service may have provided them with valuable skills, employers often prioritize candidates with civilian work experience. This lack of experience can be a significant barrier for veterans, particularly when competing against candidates who have already accumulated years of civilian job history. To address this challenge, veterans can leverage their military service by highlighting specific accomplishments, responsibilities, and leadership roles to demonstrate their ability to excel in a civilian work environment.

During their military careers, veterans dedicate their time and energy to serving their country, focusing primarily on their roles and responsibilities within the military. As a result, their exposure to civilian work environments and opportunities for professional growth outside the military sphere may be limited. This lack of civilian work experience can pose challenges when veterans transition to the civilian job market, as employers often prioritize candidates with demonstrated experience in civilian roles.

Veterans transitioning from military service to civilian life often face the challenge of limited civilian work experience. The dedication and commitment required to serve in the military often mean that veterans prioritize their military roles and responsibilities, leaving little opportunity for significant civilian work exposure. Military training and deployments consume a significant portion of veterans' careers. The rigorous training and preparation required for combat readiness leave little time for veterans to pursue civilian work experience or engage in other professional development opportunities. The intensity of military operations and the need to prioritize the demands of service contribute to veterans' primary focus on their military roles, resulting in limited exposure to civilian work environments.

Deployments and training are integral components of military service that significantly shape veterans' careers. While deployments and training provide unique and valuable experiences, they often contribute to a lack of civilian work exposure. Deployments and training are essential for ensuring military readiness and operational effectiveness. During these periods, veterans devote their time and energy to preparing for and executing missions, leaving little opportunity for meaningful civilian work experience. The demanding and unpredictable nature of military deployments can result in extended periods away from civilian job opportunities, further contributing to the gap in civilian work experience.

Military service involves rigorous and specialized training tailored to specific occupational specialties. While these training programs develop technical expertise, leadership skills, and operational proficiency, they often have limited applicability in civilian work settings. The intensive focus on military-specific training can inadvertently limit veterans' exposure to broader civilian work experiences, leading to a lack of familiarity with civilian industry practices and norms.

The involvement in deployments and training is pivotal to the military careers of veterans, granting them distinctive experiences and proficiencies. While these activities may initially result in a lack of civilian work experience, they can also serve as catalysts for veterans to acquire valuable skills and perspectives applicable to the civilian job market.

By engaging in deployments and training, veterans are introduced to demanding and ever-changing environments, nurturing the acquisition of diverse and adaptable skill sets. These skills, including leadership, teamwork, problem-solving, adaptability, and resilience, are highly sought after in the civilian workplace. The experiences gained during deployments and training can provide veterans with a solid foundation of skills that can be effectively translated and applied to various civilian work settings.

Veterans often face high-pressure situations during deployments and training, where they must make critical decisions under demanding circumstances. This exposure enhances their ability to think critically, analyze complex problems, and make sound judgments, which are skills highly valued in civilian professions. The ability to effectively manage crises and navigate uncertainty positions veterans as valuable assets to organizations seeking individuals who can thrive in fast-paced and unpredictable environments.

Deployments frequently involve interactions with diverse cultures and populations, exposing veterans to different perspectives, languages, and customs. This experience fosters cross-cultural competence, enabling veterans to work effectively with individuals from various backgrounds. In an increasingly globalized and interconnected world, the ability to navigate cultural differences and promote inclusivity is a valuable asset in the civilian work environment.

Military deployments and training often require specialized technical skills, such as operating advanced equipment, utilizing advanced technologies, or managing complex systems. These technical proficiencies, acquired through intensive training and hands-on experience, can be directly applicable to specific civilian industries or roles. Veterans possess technical expertise that can be leveraged to contribute to the growth and innovation of civilian organizations.

The military places a strong emphasis on leadership development and teamwork. Veterans emerge from deployments and training with well-honed leadership skills, having led teams in high-stakes situations. The ability to motivate, inspire, and coordinate teams is highly transferrable to civilian leadership roles. Veterans understand the value of collaboration, cooperation, and collective success, making them valuable assets to civilian organizations striving for effective teamwork and synergy.

Deployments and training instilled in veterans an adaptable and resilient mindset, enabling them to navigate change, overcome obstacles, and thrive in dynamic environments. This adaptability and resilience translate to the civilian work context, where unforeseen challenges, shifting priorities, and evolving circumstances are common. Veterans' ability to remain composed under pressure and quickly adapt to new situations positions them as valuable contributors to organizational success.

The military recognizes the importance of professional development and provides opportunities for veterans to enhance their skills and knowledge. Programs such as military education, certifications, and vocational training equip veterans with specific expertise that can be directly applied in civilian work settings. By leveraging these opportunities, veterans can bridge the gap between their military experience and civilian work experience, further enhancing their employability.

Although deployments and training initially focus on military objectives, they significantly contribute to veterans' acquisition of transferable skills, technical proficiency, leadership abilities, cross-cultural competence, and adaptability. These experiences, combined with veterans' strong work ethic and sense of mission, position them as valuable assets in the civilian job market.

The military's hierarchical structure and clearly defined career paths often limit opportunities for veterans to explore diverse occupational fields. Unlike civilians who can switch careers or gain experience in different industries, veterans may spend their entire military careers focused on a single occupational specialty or a limited range of related roles. This restricted exposure to different professional paths can result in a lack of diverse civilian work experience, making it challenging for veterans to transition smoothly into new industries.

Deployments and training predominantly take place within military installations and operational settings, which are vastly different from civilian work environments. These military settings operate under unique structures, protocols, and hierarchies that may not align with civilian workplaces. Consequently, veterans may have limited exposure to the dynamics, culture, and expectations of civilian organizations, which can pose challenges when seeking civilian employment.

While deployments and training contribute to veterans' limited civilian work experience, several strategies can help bridge the gap. Veterans should conduct a comprehensive self-assessment to identify the transferable skills acquired during their military service. This includes leadership, teamwork, problem-solving, adaptability, and resilience. Recognizing these skills allows veterans to effectively communicate their value to potential civilian employers.

The unique nature of military work often requires specialized skill sets that may not have direct civilian equivalents. Veterans develop technical expertise, leadership abilities, and operational proficiency in their specific military occupational specialties. While these skills are highly valuable and transferrable to civilian employment, they may not align directly with the requirements of civilian job descriptions. The emphasis on mastering military-specific skills and knowledge can inadvertently limit veterans' exposure to broader civilian work experiences.

Veterans possess specialized skill sets acquired through their military service, which can be highly valuable but also contribute to a lack of civilian work experience. The unique nature of military training and operations often results in veterans focusing on mastering their military roles, leaving little opportunity for significant exposure to civilian work environments.

Veterans undergo rigorous and specialized training tailored to their military occupational specialties. This training equips them with technical expertise, operational proficiency, and leadership abilities specific to their roles within the military. However, the focused nature of military training may limit exposure to the broader range of skills and experiences typically gained in civilian work settings. As a result, veterans may face challenges in areas where their military training does not directly align with civilian job requirements.

Some military roles have no direct civilian equivalents, making it challenging for veterans to translate their specialized skills into the civilian job market. The technical expertise and certifications obtained in military occupations may not have direct civilian counterparts or may require additional civilian credentials or qualifications. This lack of direct equivalence can create a barrier for veterans when seeking employment opportunities outside of their specific military occupational specialties.

While veterans may possess highly specialized skills and knowledge within their military fields, they may have limited exposure to the practices and norms of civilian industries. Military operations often function under unique structures, protocols, and hierarchies that differ significantly from civilian workplaces. As a result, veterans may have limited familiarity with industry-specific processes, technologies, and terminology, which can hinder their transition to civilian employment and result in a lack of relevant civilian work experience.

Employers who are unfamiliar with the military may perceive veterans' specialized military skill sets as having limited applicability to civilian work settings. This perception can create a barrier for veterans when competing for civilian job opportunities, as employers may undervalue or overlook the transferable skills acquired during military service. Overcoming this perception and effectively communicating the value and applicability of their specialized skills becomes crucial for veterans seeking to bridge the gap in civilian work experience.

To address the challenges associated with their specialized military skill sets and limited civilian work experience, veterans can employ several strategies. Veterans should identify the transferable skills they have developed through their specialized military training. Skills such as leadership, teamwork, problem-solving, adaptability, and critical thinking are highly valued in civilian workplaces. Recognizing and articulating these transferable skills helps veterans highlight their abilities to potential employers and demonstrates their potential for success in diverse civilian work environments.

Veterans should focus on translating their specialized military skill sets into civilian-friendly language. By highlighting relevant experiences and accomplishments using civilian terminology, veterans can effectively communicate their capabilities and align them with the requirements of civilian job descriptions. This translation helps employers understand the value and applicability of veterans' specialized skills in a civilian context.

Veterans can pursue additional professional-development opportunities, certifications, or credentials that enhance their marketability in civilian industries. This may involve obtaining civilian-equivalent qualifications or acquiring industry-specific certifications that demonstrate their proficiency and commitment to their chosen career paths outside of the military.

Veterans are driven by a deep sense of mission and service to their country. Their dedication to protecting and serving others often means that civilian work experience becomes secondary to their priorities. The selfless commitment to the military's core values, such as duty, honor, and loyalty, may lead veterans to place less emphasis on pursuing civilian work opportunities. This focused dedication to their military service can inadvertently result in limited exposure to civilian work environments and hinder the accumulation of civilian work experience.

The sense of mission and service that veterans possess instills within them a profound commitment and a strong work ethic. They are accustomed to working diligently, adhering to strict standards, and going above and beyond to accomplish objectives. This dedication translates into their civilian careers, where employers value individuals who demonstrate reliability, perseverance, and a strong sense of responsibility. Veterans' commitment and work ethic make them highly sought after by employers looking for employees who are dedicated to delivering exceptional results.

Military training and experience foster a goal-oriented mindset in veterans. They are accustomed to setting clear objectives, creating actionable plans, and relentlessly pursuing their goals. This mindset translates well into the civilian workplace, where setting and achieving goals is a fundamental aspect of professional success. Veterans' ability to align their efforts with organizational goals and maintain focus amidst challenges makes them valuable assets in achieving desired outcomes.

Integrity and professionalism are core values instilled in veterans throughout their military service. They understand the importance of ethics, discipline, and respect for rules and regulations. Veterans bring a high level of integrity and professionalism to the civilian workplace, ensuring a commitment to ethical conduct and accountability and maintaining a positive work environment. Employers value these qualities, as they contribute to organizational culture, customer trust, and long-term success.

Veterans possess a strong sense of mission, driven by a desire to serve a greater purpose beyond themselves. This mission-driven approach is a powerful motivator that fuels their dedication and commitment to their civilian careers. Veterans often seek employment opportunities that align with their values and allow them to make a meaningful impact. Their mission-driven mindset brings a sense of purpose to the workplace, inspiring their colleagues and driving positive change within organizations.

Military service exposes veterans to challenging and demanding situations that require resilience and perseverance. They have experienced adversity, overcome obstacles, and developed a mental fortitude that enables them to navigate difficult circumstances. In the civilian workforce, where setbacks and obstacles are common, veterans' resilience and perseverance allow them to maintain a positive attitude, bounce back from failures, and keep pushing forward. This determination and resilience make veterans valuable employees who can tackle challenges with tenacity and find innovative solutions.

Military service demands extensive time and energy, leaving little room for veterans to engage in professional development activities beyond their military roles. While the military provides training and educational opportunities, they are often specific to military functions and may not directly translate to civilian work settings. The lack of time and resources allocated to professional development beyond military requirements can result in veterans having limited exposure to civilian work experiences, making it challenging to compete with candidates who have accumulated civilian work experience.

Veterans often face the challenge of limited time for professional development while serving in the military. The demands of military service, including deployments, training, and operational commitments, leave little opportunity for veterans to gain extensive civilian work experience.

Military service is characterized by rigorous training, frequent deployments, and demanding schedules. The responsibilities and obligations of serving in the military leave little time for veterans to pursue civilian work experience while actively serving. Military duties often consume the majority of their time and energy, limiting opportunities to engage in professional development activities or gain exposure to civilian work environments.

Military assignments are governed by strict schedules and deployment cycles, leaving little room for flexibility in pursuing civilian work experience. Veterans may have limited control over their assignments and the locations where they serve, which restricts their ability to engage in long-term civilian internships, part-time jobs, or other opportunities that contribute to their civilian work experience. The lack of flexibility poses a significant challenge in acquiring the necessary practical knowledge and skills for the civilian job market.

Veterans' primary work environment is the military, where the culture, structure, and protocols differ significantly from civilian workplaces. While military training instills discipline, leadership, and teamwork, it may not provide sufficient exposure to the specific practices, technologies, and work dynamics of civilian industries. This limited exposure hinders veterans' understanding of civilian work environments and can make their transition to civilian careers more challenging.

Limited time for professional development often translates into limited opportunities for veterans to gain and demonstrate transferable skills that are valued in civilian workplaces. While veterans possess a wide range of skills acquired through military training, translating and showcasing these skills in a civilian context becomes a hurdle. Employers may have difficulty understanding how veterans' military experience translates to specific job requirements, leading to potential underestimation or underutilization of their skill sets.

Despite the constraints of limited time for professional development, veterans can employ various strategies to bridge the gap in civilian work experience. Veterans can make the most of the training programs and courses offered by the military that align with civilian work requirements. By selecting relevant courses, acquiring certifications, and seeking additional training opportunities, veterans can enhance their knowledge and skills in areas valued by civilian employers.

To address the limited civilian work experience resulting from the primacy of military service, veterans can take proactive steps during their transition. They should identify and acknowledge the transferable skills they have gained during their military service that are relevant to civilian employment. These skills encompass leadership, teamwork, problem-solving, adaptability, and discipline. By understanding the value of these skills in the civilian context, veterans can effectively convey their abilities to potential employers.

Veterans need to translate their military experience into civilian terms and concepts. This involves rephrasing job titles, describing responsibilities, and highlighting achievements using language familiar to civilian employers. By effectively communicating their accomplishments in a way that resonates with civilian hiring managers, veterans can bridge the gap between their military and civilian work experiences.

The nature of military work is distinct and differs from many civilian occupations. Military roles often involve specialized training, responsibilities, and environments that may not directly translate to equivalent civilian positions. Employers may have difficulty understanding how military skills and experiences can be applied in their specific industries, potentially leading to misconceptions or undervaluing of veterans' abilities. This lack of understanding further contributes to the perception that veterans lack the necessary civilian work experience.

Military work and civilian work are two distinct domains, each with its unique characteristics and demands. The nature of military service involves a strong sense of duty, hierarchical structure, specialized training, and adherence to strict regulations, while civilian work encompasses a broader range of industries, organizational structures, and professional roles. Military work is inherently driven by a mission to protect national security, defend interests, and uphold the values of the nation. Service members are bound by a sense of duty and selflessness, prioritizing the collective mission over individual objectives. The military's hierarchical structure and chain of command emphasize the importance of discipline, obedience, and unity to achieve mission success.

In contrast, civilian work is driven by organizational goals, profitability, and the provision of goods or services. The focus may vary depending on the industry and the specific organization, but the primary objective is often centered on meeting customer needs, maximizing efficiency, and driving business growth. While the sense of purpose may not be as singular and all-encompassing as in the military, civilian work still offers opportunities for individuals to find meaning and contribute to the success of their respective organizations.

In civilian work, organizational structures can vary significantly from the military culture. While hierarchies may exist, they are often more fluid and adaptable to the needs of the organization. Many civilian workplaces emphasize collaboration, teamwork, and open communication, enabling employees to contribute their expertise and ideas across different levels of the organization. The level of formality and rigidity in organizational structures can vary widely depending on the industry and company culture.

Military work involves rigorous and specialized training that prepares service members for a wide range of tasks and responsibilities. The military invests heavily in training programs to develop technical expertise, leadership skills, and operational readiness. Service members acquire specialized knowledge and skills related to their specific roles, such as combat tactics, logistics, engineering, health care, or intelligence.

In civilian work, training and skill specialization is typically more industry specific. Employees undergo training programs and educational pathways that equip them with the knowledge and skills required for their respective fields. While some roles may require specialized expertise, such as engineering, health care, or IT, many civilian positions emphasize a broader skill set that includes critical thinking, problem-solving, communication, and adaptability.

Military work often involves high operational tempo and the potential for deployment to various locations, both domestically and internationally. Service members may experience extended periods away from home and face significant physical and emotional demands during deployments. The nature of military work necessitates the ability to adapt quickly to changing environments and the willingness to face challenging and dangerous situations.

In contrast, civilian work generally follows a more predictable schedule and does not involve the same level of deployment or operational tempo. While some industries may require travel or remote work, the overall demands and potential risks are typically less intense than those experienced in military service. Civilian work allows for a greater degree of stability and work-life balance.

Understanding the unique nature of military work versus civilian work is crucial in appreciating the contrasting environments and expectations that shape these domains. The mission-oriented nature of military service, hierarchical structure, specialized training, and high operational tempo set it apart from the broader landscape of the civilian workforce environment.

When applying for civilian positions, veterans often find themselves competing against candidates who have accumulated years of experience in their respective fields. This experience can make it challenging for veterans to stand out and be seen as viable candidates, especially when job descriptions specify a certain number of years of civilian work experience. The lack of direct civilian work experience may result in veterans being overlooked, despite possessing valuable transferable skills gained during their military service.

Veterans can differentiate themselves by emphasizing their accomplishments and results achieved during their military service. Whether it's leading a team, managing complex operations, or successfully executing missions, veterans can showcase their track record of achieving tangible outcomes in challenging environments. By quantifying and articulating their achievements, veterans can demonstrate their ability to deliver results and contribute to the success of an organization, thus enhancing their competitiveness against experienced civilian candidates.

To bridge the experience gap between veterans and experienced civilian candidates, veterans should emphasize their commitment to ongoing learning and professional development. While veterans may not have the same level of civilian work experience, they can showcase their willingness to acquire new skills, pursue relevant certifications, and engage in industry-specific training programs. By demonstrating a proactive approach to learning and growth, veterans can alleviate concerns about their lack of specific experience and showcase their potential to quickly adapt and contribute in a civilian work environment.

To effectively compete with experienced civilian candidates, veterans should tailor their resumes and cover letters to highlight the relevant skills and experiences that align with the job requirements. Emphasizing transferrable skills, leadership roles, and specific achievements during military service can capture the attention of employers and showcase the unique value veterans bring to the table. Customization helps ensure that veterans' applications stand out and communicate their potential as competitive candidates.

While competing with experienced civilian candidates can pose challenges for veterans, a strategic approach can help them effectively navigate the job market and showcase their unique value. By highlighting transferrable skills, emphasizing accomplishments, investing in continued learning, building professional networks, customizing applications, and utilizing available transition programs, veterans can successfully compete for job opportunities and demonstrate their potential to excel in civilian careers. The resilience, discipline, and leadership cultivated through military service provide a platform in which veterans market and utilize to their advantage to enter into the civilian employment sector.

Seeking internships, apprenticeships, or entry-level positions can be a valuable approach for veterans to gain civilian work experience and bridge the gap. These opportunities allow veterans to learn industry-specific skills, build professional networks, and demonstrate their commitment to growth and development. While these positions may not be at the same level as their military roles, they serve as stepping stones toward securing more advanced positions in their chosen career paths.

Internships and apprenticeships serve as powerful tools for veterans to enhance their employment opportunities by gaining practical experience, expanding their skill sets, and establishing valuable connections in civilian industries. These programs provide veterans with a bridge between their military background and the civilian workforce, allowing them to develop industry-specific skills and demonstrate their potential to prospective employers. Internships and apprenticeships offer veterans the opportunity to develop practical skills that are directly applicable to their desired fields. These programs provide hands-on experience and mentorship from industry professionals, allowing veterans to apply their existing knowledge and learn new techniques and practices specific to the civilian work environment. By actively engaging in these experiences, veterans can acquire valuable skills that are in demand by employers, thereby increasing their attractiveness as candidates for employment opportunities.

Veterans can seize a unique opportunity through internships and apprenticeships to acquire firsthand exposure to various industries and establish meaningful connections in their chosen fields. These programs allow veterans to interact with professionals, learn about industry trends, and build relationships that can be beneficial for future job searches. By networking with colleagues, mentors, and supervisors during their internships or apprenticeships, veterans can tap into valuable resources, receive guidance, and access hidden job opportunities that may not be readily available through traditional channels.

Participating in internships and apprenticeships enables veterans to enhance their resumes by showcasing relevant work experience in civilian settings. While veterans may have extensive military experience, internships and apprenticeships provide tangible evidence of their ability to adapt to new environments, learn quickly, and apply their skills in a civilian context. By including these experiences on their resumes, veterans can effectively demonstrate their versatility, transferable skills, and commitment to professional growth, making them more competitive in the job market.

Internships and apprenticeships provide veterans with an opportunity to establish professional references from individuals within their desired industries. Positive recommendations and endorsements from supervisors and mentors can carry significant weight when employers assess candidates. These references serve as a testament to veterans' work ethic, skills, and professionalism, reinforcing their credibility and increasing their chances of securing future employment opportunities.

Veterans can leverage internships and apprenticeships to explore a range of career paths and gain clarity about their long-term professional objectives. These programs provide firsthand exposure to different job roles, organizational cultures, and industry-specific challenges. By actively engaging in these experiences, veterans can assess their interests, strengths, and preferences, enabling them to make informed decisions about their career trajectories. This exploration can be particularly beneficial for veterans who may be unsure about which civilian career path to pursue after their military service.

Internships and apprenticeships often provide structured transition assistance programs specifically designed for veterans. These programs aim to assist veterans in translating their military skills and experiences into civilian terms, identifying transferable skills, and aligning them with industry needs. The structured support and mentorship available through these programs can help veterans overcome the challenges of transitioning into the civilian workforce and provide guidance on how to effectively leverage their military background in a professional setting.

Veterans can seize invaluable opportunities through internships and apprenticeships to enhance their employment prospects and achieve a successful transition into civilian careers. By actively engaging in these programs, veterans can develop practical skills, expand their professional networks, enhance their resumes, and gain clarity about their career goals. Internships and apprenticeships serve as stepping stones for veterans, enabling them to bridge the gap between their military background and civilian work environments, ultimately positioning themselves as competitive candidates for employment.

3. Misunderstanding of Military Skills and Training: Another struggle veterans encounter is the misunderstanding or undervaluation of their military skills and training by civilian employers. Despite possessing specialized technical skills and undergoing extensive training, veterans may face skepticism or a lack of recognition from employers who are unfamiliar with the military. This lack of understanding can result in missed opportunities for veterans and hinder their ability to secure suitable employment. To overcome this challenge, veterans can actively seek out employers who have a track record of hiring and supporting veterans or work with organizations and resources that bridge the gap between military skills and civilian job requirements.

One of the primary reasons for misunderstanding veterans' skills is the complexity and specialization of their military training. The military operates in highly dynamic and demanding environments where veterans acquire a diverse set of technical, leadership, and problem-solving skills. However, these skills may not have direct civilian equivalents or may be expressed using different terminology. Employers, who may lack familiarity with military operations, may struggle to comprehend the depth and applicability of these skills. To address this, veterans should effectively translate their military experiences into civilian terms and highlight the transferable skills that align with specific job requirements.

Veterans often possess nontraditional education and certification pathways. While they may not have obtained formal degrees or diplomas, veterans often undergo intensive and comprehensive training in their respective military occupational specialties. Unfortunately, some employers may prioritize traditional academic credentials, overlooking the substantial knowledge and competence acquired through military training. Veterans can counter this misunderstanding by clearly articulating the depth and quality of their military education and certifications, emphasizing the practical application and real-world experience gained during their service.

The military culture, with its distinct values, hierarchical structure, and communication norms, can create a barrier to understanding veterans' skills and experiences. The strict adherence to protocols, disciplined work environment, and efficient decision-making processes in the military may be perceived as rigid or inflexible by civilian employers. Additionally, veterans may struggle to adapt their communication style, which tends to be direct and concise, to the more nuanced and collaborative approach often favored in civilian workplaces. By raising awareness about these cultural differences and proactively adapting their communication style, veterans can help employers better understand and appreciate their skills and training.

Bias and stereotypes can significantly contribute to the misunderstanding of veterans' skills and training. Some employers may hold preconceived notions that veterans are solely suited for certain roles, such as security or physical labor, and may overlook their potential to excel in diverse fields. Combat-related experiences may also lead to misconceptions about veterans' emotional well-being or ability to adapt to civilian work environments. To counter these biases, veterans can leverage networking opportunities; participate in professional organizations; and engage in informational interviews to showcase their skills, challenge stereotypes, and educate employers about the broad range of talents they bring to the table.

Addressing the misunderstanding of veterans' skills and training requires collaborative efforts between veterans, employers, and support organizations. Employers can benefit from partnering with veteran-focused organizations, participating in career fairs, and providing targeted training programs that bridge the gap between military and civilian skills. Veterans can also proactively seek out resources and programs that facilitate their transition, such as skills-translation workshops and mentorship initiatives. By fostering these collaborative relationships, both veterans and employers can enhance their understanding of each other's needs and create a mutually beneficial work environment.

4. Postservice Mental Health Challenges: Transitioning out of the military can be emotionally and mentally challenging for veterans, and the impact of these challenges can affect their job search. Many veterans experience mental health issues such as PTSD, depression, or anxiety as a result of their service. These conditions can affect their ability to interview confidently, adapt to new work environments, or effectively communicate their skills and experiences to potential employers. Addressing mental health challenges is crucial for veterans' well-being and their ability to navigate the job market successfully. Providing accessible mental health resources and support networks is essential for helping veterans overcome these struggles.

PTSD is a common mental health condition among veterans, resulting from exposure to traumatic events during military service. PTSD can manifest in intrusive thoughts, flashbacks, hypervigilance, and avoidance behaviors that can significantly impact daily functioning, including employment. Symptoms of PTSD may affect veterans' ability to concentrate, manage stress, and effectively interact with colleagues or employers. Employers can support veterans by fostering a workplace culture that promotes understanding and provides accommodations, such as flexible work hours or confidential mental health resources.

Depression and anxiety are prevalent mental health conditions among veterans. The challenges associated with transitioning to civilian life, including uncertainty, loss of camaraderie, and difficulty adjusting to new routines, can exacerbate these conditions. Depression and anxiety can affect motivation, energy levels, and social interactions, potentially impeding veterans' pursuit of employment opportunities. Employers can create a supportive environment by offering mental health resources, implementing employee-assistance programs, and fostering open communication channels to address these challenges.

Veterans may face an increased risk of substance abuse disorders, often as a coping mechanism for managing the emotional and psychological stressors they encounter. Substance abuse can have severe consequences on veterans' mental and physical health, leading to impaired decision-making, strained relationships, and challenges in maintaining employment stability. Employers can play a crucial role in supporting veterans' recovery by providing access to substance-abuse treatment programs, offering flexible schedules to accommodate therapy or support-group attendance, and promoting a workplace culture that is understanding and free from stigma.

Traumatic Brain Injury (TBI) is another common issue faced by veterans, often resulting from combat-related incidents or accidents during service. TBI can lead to cognitive impairments, memory deficits, and difficulties with concentration and problem-solving. These challenges can pose barriers to securing employment or performing job-related tasks effectively. Employers can support veterans with TBI by providing reasonable accommodations, such as assistive technologies, modified workstations, or additional training opportunities that help optimize their work performance.

The stigma surrounding mental health remains a significant barrier for veterans seeking employment. Many veterans may hesitate to disclose their mental health challenges due to concerns about potential discrimination or negative perceptions. Fear of stigma can prevent veterans from accessing the support they need and contribute to a reluctance to pursue employment opportunities. Employers can combat stigma by fostering an inclusive and nonjudgmental work environment, implementing diversity and inclusion initiatives, and promoting mental health awareness and education among their workforce.

Access to mental health resources is critical for veterans facing postservice mental health challenges. Unfortunately, many veterans encounter barriers to accessing quality mental health care, such as long wait times, limited availability of specialized services, or lack of awareness about available resources. Employers can collaborate with health-care providers and community organizations to ensure veterans have access to comprehensive mental health support, including counseling, therapy, and peer support networks.

5. Networking and Professional Connections: Veterans often face difficulties in building and leveraging professional networks and connections when transitioning to civilian employment. Networking plays a crucial role in the job search process, as many job opportunities are filled through personal referrals and recommendations. However, veterans may not have extensive civilian networks or may feel unsure about how to navigate the civilian professional landscape. Encouraging veterans to participate in networking events, connecting with veteran-specific organizations and professional associations, and engaging in mentorship programs can help veterans expand their professional connections and increase their chances of finding suitable employment opportunities.

Numerous veteran service organizations and career resources are dedicated to supporting veterans during their transition to civilian employment. These organizations provide networking events, job fairs, career counseling, and resume-building workshops tailored specifically to veterans' needs. Engaging with these resources enables veterans to expand their professional connections, gain industry-specific knowledge, and access employment opportunities that cater to their unique skill sets and experiences.

Joining professional associations and attending industry events can significantly enhance veterans' networking capabilities. These associations provide a platform for veterans to connect with professionals in their desired fields, learn about industry trends, and gain exposure to job opportunities. Active participation in industry events, conferences, and workshops allows veterans to showcase their expertise, establish relationships with key stakeholders, and stay updated on the latest advancements in their chosen industries.

In today's digital age, online networking platforms offer a convenient and efficient way for veterans to connect with professionals across various industries. Platforms such as LinkedIn provide veterans with a virtual space to showcase their skills, experiences, and achievements while also enabling them to connect with recruiters, industry experts, and potential employers. Active engagement on these platforms, including participating in industry-specific groups, sharing relevant content, and reaching out to professionals, can lead to valuable connections and employment opportunities.

Informational interviews and job shadowing experiences allow veterans to gain insights into different industries, job roles, and company cultures. By reaching out to professionals in their desired fields, veterans can request informational interviews to learn about specific career paths, seek advice, and expand their networks. Additionally, job shadowing opportunities provide firsthand exposure to the day-to-day activities and requirements of various roles, helping veterans make informed decisions about their career paths and potentially leading to employment opportunities.

Networking and professional connections are pivotal for veterans as they transition out of the military and pursue civilian employment. By leveraging their military networks, engaging with veteran service organizations, participating in industry events, seeking mentorship, utilizing online platforms, and exploring informational interviews, veterans can expand their professional connections, access job opportunities, and gain invaluable guidance and support. Networking empowers veterans to showcase their unique skills and experiences, increase their visibility in the job market, and ultimately achieve successful employment transitions.

6. Geographic and Industry-Specific Challenges: The availability of employment opportunities can vary significantly depending

> on geographic location and industry. Veterans who have specific skills or experience that may not align with the local job market or who are transitioning into industries where they have limited knowledge or exposure may face additional challenges in securing employment. Overcoming these challenges may involve exploring relocation options, gaining additional education or certifications, or seeking out employers who prioritize the unique skill sets that veterans possess. In some cases, veterans may need to consider exploring relocation options to areas where their skills are in high demand.

As veterans transition from military service to civilian life, they often encounter geographic and industry-specific challenges that can impact their ability to secure suitable employment opportunities. These challenges arise due to factors such as geographic limitations, unfamiliarity with local job markets, and the need to bridge the gap between military skills and industry requirements. Veterans may face geographic challenges when seeking employment due to factors such as family obligations, limited financial resources, or a desire to remain near their military support networks. Relocation constraints can limit the pool of available job opportunities and require veterans to adapt their job search strategies accordingly.

Transitioning to a new geographic location often means veterans must familiarize themselves with the local job market dynamics, including industry trends, employment opportunities, and networking resources. Lack of awareness about the local job market can hinder veterans' ability to effectively target their job search efforts and connect with relevant employers and industry professionals.

Veterans often face industry-specific challenges when transitioning into civilian employment. One prominent hurdle is the need to translate their military skills into industry-specific terms. While veterans possess a distinct skill set acquired through their military service, civilian employers may struggle to fully grasp the relevance and applicability of these skills. As a result, there is a risk of underestimating veterans' qualifications. To overcome this obstacle, veterans must effectively communicate and translate their military experiences and skills into the language and requirements of their targeted industries. By clearly articulating how their military expertise aligns with the demands of civilian roles, veterans can enhance their chances of being recognized and valued by employers in their desired industries. This process of bridging the gap between military and civilian terminology is essential for veterans to successfully navigate industry-specific challenges during their career transition.

Veterans transitioning into civilian employment often encounter industry-specific challenges that can impact their job prospects. One such challenge is the existence of training and certification requirements in certain industries that veterans may not have fulfilled during their military service. This can place them at a disadvantage when competing against civilian candidates who possess the necessary credentials. To overcome this obstacle, veterans can take proactive measures by pursuing relevant certifications, enrolling in vocational training programs, or highlighting how their military training has equipped them with the aptitude and capacity to acquire new skills.

Another challenge that veterans may face is industry bias and stereotypes. Misconceptions about military culture, leadership styles, or assumptions about veterans' adaptability to civilian work environments can create barriers to employment. To counteract these biases, employers and industry professionals must be educated about the valuable attributes that veterans bring to the table. By raising awareness of the discipline, adaptability, and teamwork ingrained in veterans, these biases can be challenged, fostering a more inclusive and supportive environment for veterans in the industry.

Navigating these industry-specific challenges requires veterans to be proactive, resilient, and resourceful. By actively pursuing relevant certifications, debunking stereotypes, and showcasing their transferable skills, veterans can overcome these hurdles and successfully transition into meaningful careers in their desired industries.

To overcome the challenges posed by geographic and industry-specific factors, veterans can employ several strategies. Firstly, conducting comprehensive research on the local job market is crucial. This includes gaining insights into key industries, in-demand skills, and networking opportunities. By engaging with industry professionals, attending career fairs, joining industry associations, and utilizing online networking platforms, veterans can expand their professional connections and stay informed about industry trends and opportunities.

Another effective strategy is tailoring resumes and cover letters to the targeted industry. Veterans should customize these documents to highlight their transferable skills and experiences that are directly relevant to the industry they are pursuing. Emphasizing accomplishments, leadership roles, and specific achievements during their military service can effectively demonstrate their value to potential employers and increase their chances of standing out in the competitive job market.

Combining thorough research, active networking, and targeted customization of application materials, veterans can proactively address the challenges posed by geographic and industry-specific factors. These strategies can help veterans position themselves effectively, showcase their qualifications, and enhance their opportunities for success in their desired industries.

To enhance their career prospects and overcome industry-specific challenges, veterans should prioritize professional development and training. Actively pursuing opportunities such as industry-specific training programs, vocational courses, or online certifications showcases their commitment to acquiring the necessary skills valued by employers. By investing in continuous learning and staying updated on industry trends, veterans can bolster their marketability and adaptability in their targeted industries.

Additionally, engaging with veteran mentorship programs and utilizing transition assistance services can be immensely beneficial. These resources provide guidance and support tailored to veterans' needs during their transition into civilian employment. Mentors, who have firsthand experience in the industry, can offer valuable insights, share their career journeys, and provide guidance on navigating industry-specific challenges. Transition assistance services can provide practical support in terms of resume building, interview preparation, and connecting veterans with industry-specific opportunities and employers.

By actively participating in professional development initiatives and seeking mentorship and transition assistance, veterans can equip themselves with the necessary skills, knowledge, and networks to overcome industry-specific challenges. These proactive steps can greatly enhance their career prospects and ensure a smoother transition into their desired industries.

Employment opportunities for veterans not only provide economic stability and career growth but also play a significant role in ensuring access to essential health benefits. Veterans often face unique health challenges resulting from their military service, and securing employment that offers comprehensive health benefits becomes crucial for their overall well-being. Employment opportunities that provide comprehensive health benefits can significantly impact veterans' access to health-care services. The VA offers a range of health-care services for eligible veterans, but having additional health coverage through employment can complement and enhance the care they receive. Employer-sponsored health insurance plans can cover a broader range of services, including specialist care, preventive care, and prescription medications, ensuring veterans have access to comprehensive health-care coverage.

Securing employment that offers health benefits enables veterans to maintain continuity of care, ensuring seamless access to health-care services. Transitioning from military health-care systems to civilian providers can be challenging; but with employer-sponsored health benefits, veterans can establish relationships with health-care providers, access consistent medical care, and receive ongoing treatment for any service-related health issues. Continuity of care plays a crucial role in managing chronic conditions, mental health concerns, and other health needs veterans may have.

Veterans' well-being is significantly enhanced by employment opportunities that prioritize mental health benefits. Many veterans experience mental health conditions such as PTSD, anxiety, and depression as a result of their military service. Access to mental health services, including therapy and counseling, through employer-sponsored health benefits, can help veterans address and manage these challenges effectively. Timely and consistent mental health support promotes resilience, reduces stigma, and enhances veterans' overall quality of life.

Veterans with service-connected disabilities often require ongoing medical care, rehabilitation services, and assistive devices. Employment opportunities that offer health benefits ensure that veterans with disabilities can access the necessary support. Comprehensive health coverage can cover expenses related to medical equipment, specialized therapies, and adaptive technologies, easing the financial burden on veterans and enabling them to lead independent and fulfilling lives.

Employment opportunities with health benefits also facilitate access to preventive care and wellness programs. Veterans, like the general population, benefit from routine screenings, vaccinations, and health promotion activities that help prevent or detect health conditions early on. Employer-sponsored wellness programs can provide resources for maintaining a healthy lifestyle, managing chronic conditions, and promoting overall wellness. These initiatives empower veterans to proactively manage their health, prevent complications, and enhance their quality of life.

Health benefits offered through employment opportunities extend beyond the veteran themselves, often providing coverage for their dependents as well. Dependents, including spouses and children, can access health-care services, specialist care, and preventive care through employer-sponsored health plans. This comprehensive coverage alleviates the burden on veterans by ensuring that their loved ones have access to necessary medical care and support.

Employment opportunities for veterans are intrinsically linked to access to health benefits, providing essential support for their physical and mental well-being. Comprehensive health-care coverage, continuity of care, mental health support, assistance for service-connected disabilities, access to preventive care, and support for dependents all contribute to veterans' overall health. By securing employment opportunities that offer comprehensive health benefits, veterans can focus on their careers while knowing that their health-care needs are met, allowing them to thrive both personally and professionally.

Chapter 9

Health Care and Benefits

When veterans transition out of military service, they are provided with a range of health care and benefits to support their physical and mental well-being. Recognizing the unique challenges and health needs veterans may face, the VA offers comprehensive health-care services, financial assistance, educational benefits, and other resources. In this chapter, we will explore the health care and benefits veterans receive when transitioning out of the service, highlighting the importance of these provisions in ensuring a smooth and successful transition.

1. Health-Care Services: (a) VA Medical Care: One of the primary health-care benefits for transitioning veterans is access to VA medical care. Veterans are eligible for a wide range of health-care services, including preventive care, primary care, specialty care, mental health services, and prescription medications. The VA health-care system offers comprehensive coverage to address both service-connected and nonservice-connected conditions, ensuring that veterans receive the necessary care to maintain their health and well-being.

One of the foundational aspects of the VA health-care system is primary care. Veterans have access to comprehensive primary care services, including routine check-ups, preventive care, and treatment of acute and chronic illnesses. Primary care providers within the VA system focus on promoting overall wellness, disease prevention, and early intervention to ensure veterans' health is monitored and managed effectively.

The VA health-care system offers a wide range of specialized medical services to address the unique health needs of veterans. These services encompass various medical disciplines, including but not limited to (a) mental health services: the VA places a significant emphasis on mental health, providing specialized care for conditions such as PTSD, depression, anxiety, and substance abuse. The VA offers counseling, therapy, medication management, and comprehensive mental health programs to support veterans' mental well-being.

The VA provides comprehensive health-care services for female veterans, including preventive care, maternity care, gynecological services, and reproductive healthcare. Specialized women's health clinics and programs ensure that female veterans receive gender-specific care tailored to their needs.

The VA recognizes the unique health-care requirements of aging veterans and provides specialized geriatric care services. This includes comprehensive assessment and management of age-related conditions, coordination of care, and support for elderly veterans to maintain their independence and quality of life.

Veterans with injuries or disabilities resulting from their military service can access rehabilitation and physical therapy services. These services aim to enhance mobility, restore function, and improve veterans' quality of life through individualized treatment plans and specialized therapies. The VA health-care system includes a wide range of specialist services such as cardiology, orthopedics, dermatology, ophthalmology, and more. Veterans can receive specialized consultations, diagnostic procedures, and treatment for specific medical conditions from expert health-care providers within the VA network.

The VA also provides transition assistance to veterans to help them navigate the health-care system as they transition out of the service. Programs such as the VA TAP and the Benefits Delivery at Discharge (BDD) program offer counseling, information, and resources to facilitate a smooth transition into civilian health care and benefits.

The VA operates an extensive pharmacy network, providing veterans with access to medications at no or reduced cost. Veterans can obtain prescription medications through VA pharmacies, mail-order services, or community care providers. The VA also employs robust medication-management programs, including medication reviews, education, and counseling, to ensure safe and effective medication use.

Preventive care and health promotion play a significant role in the VA health-care system. Veterans have access to a range of preventive services, including vaccinations, cancer screenings, health-risk assessments, and health-education programs. The VA emphasizes preventive care to detect and manage health conditions at an early stage, promoting overall wellness and reducing the burden of preventable diseases.

To enhance accessibility and convenience, the VA has expanded its telehealth and remote care capabilities. Veterans can access health-care services through telehealth platforms, allowing them to consult with health-care providers remotely. Telehealth services enable veterans to receive timely medical advice, follow-up care, and certain treatments without the need for in-person visits, improving access to health care, particularly for those in rural or remote areas.

2. Disability Benefits: Veterans who have incurred disabilities as a result of their military service are entitled to disability benefits. The VA provides compensation to veterans based on the degree of their service-connected disability, which can include physical injuries, mental health conditions, and other impairments. Disability benefits help veterans receive financial support to cope with the challenges posed by their disabilities and improve their quality of life.

Disability benefits are a crucial component of the comprehensive support provided to veterans by the VA. These benefits recognize the sacrifices made by veterans during their service and aim to provide financial assistance and support for service-connected disabilities. Disability benefits are primarily provided to veterans who have incurred disabilities or injuries as a result of their military service. Service-connected disabilities refer to physical or mental health conditions that can be directly attributed to military service. The VA evaluates the severity and impact of these disabilities to determine eligibility for disability benefits.

Serving in the military demands immense dedication, courage, and sacrifice. Veterans often face significant challenges during their service, which can result in service-connected disabilities. These disabilities, arising from injuries or illnesses directly related to their time in the military, can have a profound impact on veterans' lives. In recognition of their sacrifices, society has a moral obligation to support these veterans and ensure they receive the necessary care and assistance.

Service-connected disabilities encompass physical and mental health conditions resulting from military service. These disabilities can manifest during or after active duty, affecting veterans' abilities to engage in daily activities, pursue employment opportunities, or maintain stable relationships. Common service-connected disabilities include combat injuries, PTSD, TBIs, hearing loss, and other conditions related to exposure to environmental hazards, such as Agent Orange or burn pits.

Service-connected disabilities significantly impact veterans' lives, often causing physical pain, psychological distress, and reduced quality of life. These disabilities can be a constant reminder of the challenges and traumas experienced during military service. They may lead to long-term health complications, functional limitations, and difficulties in reintegrating into civilian life. The toll of service-connected disabilities extends beyond the individual veterans, affecting their families, communities, and the broader society.

Recognizing the immense sacrifices made by veterans, numerous support systems are in place to assist those with service-connected disabilities. The VA plays a crucial role in delivering comprehensive health care, disability compensation, vocational rehabilitation, and other services to eligible veterans. The VA's health-care system offers specialized care for veterans with service-connected disabilities, including access to medical specialists, rehabilitative therapies, assistive devices, and mental health counseling. Additionally, disability compensation provides financial support to veterans whose disabilities result in a diminished ability to work or perform everyday tasks.

Beyond the VA, there are various nonprofit organizations, community programs, and initiatives aimed at supporting veterans with service-connected disabilities. These organizations provide a range of services, including housing assistance, employment training, adaptive sports and recreational activities, and peer support networks. They play a vital role in complementing the efforts of the VA, ensuring veterans receive holistic support tailored to their specific needs.

While progress has been made in supporting veterans with service-connected disabilities, challenges persist. These include bureaucratic complexities, long waiting times for VA services, insufficient resources, and gaps in mental health care. Addressing these challenges requires ongoing collaboration between government agencies, nonprofit organizations, health-care providers, and the public. Increased funding, streamlined processes, enhanced outreach efforts, and improved coordination among stakeholders can help ensure that veterans receive the timely and comprehensive support they deserve.

Service-connected disabilities are a somber reminder of the sacrifices made by our veterans in the defense of our nation. It is our collective responsibility to honor their service and provide them with the necessary support and resources to lead fulfilling lives. By bolstering health-care, compensation, rehabilitation, and community-based initiatives, we can ensure that veterans with service-connected disabilities receive the care, understanding, and opportunities they need to thrive. Together, we can repay our debt to these brave men and women, standing by their side as they navigate the challenges of life after service.

Veterans with service-connected disabilities may be eligible for compensation from the VA. Compensation is a financial benefit that aims to provide support for the impact of disabilities on veterans' daily lives, earning capacity, and quality of life. The amount of compensation is determined based on the degree of disability, ranging from partial disability to total disability, and is calculated using the VA's disability rating system.

When veterans experience disabilities as a result of their military service, compensation becomes a crucial aspect of their support. Compensation aims to acknowledge the sacrifices made by veterans and provide financial assistance to mitigate the challenges posed by service-connected disabilities. Compensation for disabilities refers to financial benefits provided to veterans who have incurred physical or mental impairments during their military service.

Compensation for disabilities holds great significance for veterans and their families, providing vital financial support in the face of medical expenses, reduced earning capacity, and other challenges arising from service-connected disabilities. It serves as a tangible acknowledgment of the sacrifices made by veterans, affirming society's commitment to standing by those who have defended our nation. Compensation enables veterans to access necessary health care, assistive devices, and adaptive technologies, promoting their overall rehabilitation and enhancing their quality of life.

The VA plays a central role in determining eligibility for compensation and establishing the degree of disability for individual veterans. The VA assigns disability ratings based on the severity of impairments, ranging from 0 percent to 100 percent. These ratings consider medical evidence, clinical assessments, and other factors, such as the impact on daily activities, employability, and social functioning. A higher disability rating corresponds to increased compensation, reflecting the level of impairment experienced by veterans.

Compensation for disabilities encompasses various components designed to provide support tailored to the specific needs of veterans. These components include the following:

- Veterans with service-connected disabilities receive monthly tax-free payments based on their disability rating. The VA's rating schedule determines the corresponding compensation amount, with higher ratings resulting in increased financial support.
- Special Monthly Compensation (SMC) provides additional financial assistance to veterans with severe disabilities, such as loss of limbs, loss of sight or hearing, and need for aid and attendance. SMC recognizes the extraordinary circumstances faced by veterans with exceptionally challenging impairments.
- Concurrent Retirement and Disability Pay (CRDP) allows certain disabled veterans who are also receiving military retirement pay to receive both retirement and disability

compensation, eliminating or reducing the offset between the two benefits.

- Individual Unemployability (IU) provides additional compensation to veterans whose service-connected disabilities prevent them from maintaining substantial employment. It ensures that veterans with severe impairments receive adequate financial support, even if their disability rating does not reach 100 percent.

Compensation for disabilities is a vital aspect of supporting veterans who have sacrificed their physical and mental well-being in service to our country. It acknowledges their sacrifices, eases financial burdens, and helps veterans access necessary healthcare and support services. The VA's disability rating system, combined with various compensation mechanisms, ensures that veterans receive fair and equitable compensation corresponding to the impact of their service-connected disabilities. By providing meaningful financial support, we honor the dedication and sacrifice of our veterans, reaffirming our commitment to their well-being and recognizing their ongoing contributions to society.

The VA utilizes a disability rating system to assess the severity and impact of service-connected disabilities. This system assigns a rating percentage to each disability, ranging from 0 percent to 100 percent, in increments of 10 percent. The rating is based on the impact of the disability on the veteran's ability to function and perform daily activities. The higher the rating percentage, the greater the level of disability and the corresponding compensation.

When veterans experience service-connected disabilities, it is crucial to have a comprehensive process in place to evaluate their conditions accurately. The disability evaluation process serves as the foundation for determining the extent of impairments, eligibility for benefits, and appropriate compensation. The disability evaluation process typically begins with the veteran submitting an initial application to the VA. This application includes details about the veteran's military service, specific disabilities, and any supporting medical evidence. Veterans need to provide comprehensive and accurate information to facilitate the evaluation process.

Upon receiving the initial application, the VA schedules a medical examination for the veteran. This examination is conducted by a qualified health-care professional who assesses the severity of the claimed disabilities and determines their relationship to the veteran's military service. The medical examination plays a crucial role in establishing the foundation for the disability evaluation process.

Following the medical examination, the gathered evidence, including the examination results and any additional medical documentation, is reviewed by rating authorities within the VA. These rating authorities assess the degree of disability by comparing the veteran's impairments to the VA's Schedule for Rating Disabilities (VASRD). The VASRD provides guidelines and criteria for evaluating the severity and impact of various disabilities.

Based on the assessment, the rating authorities assign a disability rating to each service-connected condition. This rating ranges from 0 percent to 100, with higher percentages reflecting more severe impairments. The rating decision is then communicated to the veteran through a formal notification letter, detailing the assigned disability ratings and the effective date of the decision. This notification also outlines the appeal process if the veteran disagrees with the assigned ratings.

If a veteran disagrees with the assigned disability ratings, they have the right to appeal the decision. The appeals process allows veterans to present additional evidence, such as medical records or expert opinions, to support their case. The VA's Board of Veterans' Appeals (BVA) reviews the appeal and decides based on the submitted evidence and applicable laws and regulations. If necessary, veterans can pursue further appeals to higher-level review or seek judicial review through the US Court of Appeals for Veterans Claims.

The disability evaluation process is not static. The VA recognizes that service-connected disabilities may change over time. Therefore, veterans may undergo periodic reevaluations to assess the progression or improvement of their conditions. Reevaluations help ensure that veterans receive the appropriate level of compensation and support that aligns with their current medical status.

Throughout the disability evaluation process, veterans are encouraged to seek support and assistance from various resources. Veteran service organizations, accredited representatives, and legal professionals specializing in veterans' affairs can provide guidance, advocacy, and assistance in navigating the complex evaluation process. These support systems play a crucial role in ensuring that veterans receive fair and accurate evaluations of their service-connected disabilities.

The process for veterans' disability evaluation is a vital component of providing fair and appropriate support to those who have served their country. From the initial application to the rating decision and potential appeals, this process ensures that service-connected disabilities are evaluated accurately and veterans receive the necessary benefits and compensation they deserve. By ensuring transparency, fairness, and accessibility, we can honor the sacrifices of our veterans and provide them with the support they need to lead fulfilling lives after their military service.

As mentioned earlier, veterans with service-connected disabilities receive compensation to address the financial impact of their disabilities. The compensation amount varies based on the disability rating and may be adjusted over time as the condition improves, worsens, or stabilizes.

When veterans experience service-connected disabilities, various types of disability benefits are available to provide them with the necessary support and assistance. These benefits aim to recognize the sacrifices made by veterans and help mitigate the challenges posed by their disabilities.

Dependency and Indemnity Compensation (DIC) is a benefit provided to surviving spouses, children, and dependent parents of veterans who died as a result of service-connected disabilities or certain other qualifying circumstances. DIC offers a monthly monetary allowance to help ease the financial burden faced by survivors following the loss of a veteran. It recognizes the sacrifices made by the veteran and aims to provide ongoing support to their loved ones.

In addition to the aforementioned disability benefits, veterans may also be eligible for various supportive benefits, such as health-care services, vocational rehabilitation, housing assistance, and adaptive equipment. These benefits aim to address the unique needs of veterans with service-connected disabilities, promoting their overall well-being, independence, and successful reintegration into civilian life.

The availability of diverse disability benefits is a testament to society's commitment to supporting veterans with service-connected disabilities. Disability compensation, special monthly compensation, individual unemployability, concurrent retirement and disability pay, and dependency and indemnity compensation form a comprehensive system of support, providing financial assistance and acknowledging the sacrifices made by veterans and their families. By offering these benefits, we recognize the immense contributions and sacrifices of our veterans and strive to ensure they receive the support they need to lead meaningful and fulfilling lives beyond their military service.

The VA's VR&E program assists veterans with service-connected disabilities in achieving meaningful employment and gaining economic independence. Through VR&E, eligible veterans receive support and resources for vocational rehabilitation, educational training, job-placement assistance, and self-employment opportunities. The program focuses on empowering veterans with disabilities to develop new skills, transition into civilian careers, and enhance their employability.

Transitioning from military service to civilian life can present unique challenges for veterans, particularly in terms of securing meaningful employment. Vocational rehabilitation and employment programs play a vital role in supporting veterans by equipping them with the necessary skills, resources, and support to successfully reintegrate into the workforce. The VA offers a comprehensive range of vocational rehabilitation and counseling services to eligible veterans. These services provide personalized assistance to veterans with service-connected disabilities, helping them overcome barriers to employment and achieve their vocational goals. Vocational Rehabilitation and Counseling (VR&C) services encompass vocational assessment, career counseling, skills training, job placement, and on-the-job support, tailored to the specific needs and abilities of each veteran.

Education and training programs form a significant part of vocational rehabilitation and employment initiatives for veterans. The Post-9/11 GI Bill and other educational benefits provide financial assistance to veterans pursuing higher education, vocational training, apprenticeships, or certification programs. These programs enhance veterans' marketable skills, expand their career options, and increase their competitiveness in the civilian job market.

The VR&E program, also known as Chapter 31, is a key component of vocational rehabilitation and employment for veterans. It focuses on assisting veterans with significant service-connected disabilities in achieving employment or achieving greater independence in their daily lives. Through the VR&E program, veterans receive comprehensive services such as vocational evaluation, employment counseling, education or training, and assistive technology and accommodations as needed. The program aims to empower veterans to overcome their disabilities and attain meaningful and sustainable employment.

Efforts to enhance vocational rehabilitation and employment for veterans also involve collaboration with employers and job-placement support. The VA, in partnership with public and private entities, establishes relationships with employers who value the unique skills, experience, and work ethic that veterans bring to the workforce. These collaborations promote veteran-friendly hiring practices, job fairs, and networking opportunities, ensuring veterans have access to a range of employment options aligned with their abilities and aspirations.

For veterans interested in starting their businesses, entrepreneurship, and small business assistance programs are available. The Small Business Administration (SBA) offers resources, training, mentoring, and financial assistance to help veterans launch and sustain their entrepreneurial ventures. These programs recognize the potential of veterans as business owners, encouraging self-employment and fostering economic growth in local communities.

Vocational rehabilitation and employment programs are integral to ensuring the successful transition and long-term success of veterans in the civilian workforce. By providing comprehensive services, educational benefits, counseling, and job-placement support, these initiatives empower veterans to overcome challenges, leverage their unique skills and experiences, and secure meaningful and sustainable employment. The collaborative efforts between government agencies, employers, and veterans themselves ensure that the talents and contributions of veterans are recognized, leading to enriched lives, economic stability, and successful reintegration into society.

5. Appeals and Support Services: The VA recognizes the importance of ensuring fair and accurate disability determinations. Veterans who disagree with their disability ratings or have issues with their claims can pursue appeals through the VA's appeals process. Additionally, the VA provides support services such as case management, counseling, and vocational rehabilitation counseling to assist veterans in navigating the disability benefits process and accessing the support they need.

Navigating the complex landscape of benefits and services can be challenging for veterans seeking assistance from the VA. Appeals and support services provided by the VA are crucial in ensuring fairness, accountability, and comprehensive assistance for veterans.

The VA recognizes that veterans may disagree with decisions made regarding their benefits, disability ratings, or other claims. As a result, a formal appeals process is in place to provide veterans with the opportunity to challenge and seek reconsideration of unfavorable decisions. The appeals process involves several stages:

1. Notice of Disagreement (NOD): Veterans initiate the appeals process by submitting a NOD within the specified timeframe. The NOD outlines the issues being contested and requests a review.
2. Decision Review Officer (DRO) Review: After the NOD is filed, a Decision Review Officer conducts a review of the case, considering any new evidence submitted. This review provides veterans with the opportunity to present their cases and have their concerns addressed.
3. Board of Veterans' Appeals (BVA): If the veteran remains dissatisfied with the decision, they can appeal to the Board of Veterans' Appeals (BVA). The BVA conducts an independent review of the case, considering the evidence, arguments, and

applicable laws and regulations. The BVA may grant the appeal, reverse the decision, or remand the case for further evaluation.

4. Appeals to Higher Courts: If the veteran disagrees with the BVA's decision, they may seek further review by appealing to the US Court of Appeals for Veterans Claims and, if necessary, to the US Court of Appeals for the Federal Circuit. These higher courts provide a final recourse for veterans seeking resolution and justice.

In addition to the appeals process, the VA offers a range of support services to assist veterans in various aspects of their lives. These services aim to address the unique challenges faced by veterans and promote their overall well-being:

- The Veterans Board of Appeals (VBA) assists in navigating the complex benefits system, including education benefits, vocational rehabilitation, home loans, and life insurance. VBA representatives help veterans understand their entitlements, process claims, and access the benefits they have earned through their military service.
- The Veterans Health Administration (VHA) provides a comprehensive health-care system tailored to veterans' needs. It offers medical services, mental health care, specialized rehabilitation programs, and access to an extensive network of health-care professionals. The VHA ensures that veterans receive the necessary medical support to address service-related conditions and promote overall wellness.
- Vet Centers provide a wide range of counseling, outreach, and referral services to veterans and their families. These centers offer readjustment counseling for combat veterans, bereavement counseling for families of fallen service members, and counseling for veterans who have experienced military sexual trauma. Vet Centers serve as a crucial resource for veterans in need of mental health support and community connection.

Disability benefits for veterans play a critical role in recognizing and supporting those who have incurred disabilities as a result of their military service. These benefits provide financial assistance, compensation, and access to vocational rehabilitation to enhance veterans' well-being, financial stability, employability, and educational opportunities to facilitate the ultimate success in their transition to civilian life.

3. Education and Training Benefits: One of the most significant benefits for transitioning veterans is the GI Bill, which provides educational assistance to pursue higher education or vocational training. The GI Bill offers financial support for tuition, books, housing allowances, and other educational expenses, enabling veterans to acquire new skills, enhance their qualifications, and transition into civilian careers.

In recognition of the sacrifices made by veterans in serving their nations, governments and organizations have implemented a range of education and training benefits to support their transition to civilian life. These benefits are designed to provide veterans with opportunities for academic advancement, vocational training, and skill development. One of the most significant education benefits available to veterans is the Post-9/11 GI Bill. This comprehensive program offers financial assistance for education and training to veterans who served on active duty after September 10, 2001. Under this bill, veterans are eligible for funding to pursue degrees, vocational training, apprenticeships, and various certification programs. The GI Bill covers tuition and fees, a housing allowance, and a stipend for books and supplies. It also provides support for distance learning and online courses, making education accessible to veterans regardless of their location.

The VR&E Program, administered by the Department of Veterans Affairs (VA), offers comprehensive support to veterans with service-connected disabilities. Through this program, eligible veterans receive vocational counseling, assessment of their transferable skills, and assistance in developing personalized rehabilitation plans. The VR&E Program covers the cost of education or training programs, including tuition, fees, books, supplies, and even necessary adaptive equipment.

The aim is to help disabled veterans reintegrate into the workforce by acquiring new skills or enhancing existing ones.

The Yellow Ribbon Program supplements the Post-9/11 GI Bill and provides additional financial support to veterans attending private or out-of-state colleges and universities. Under this program, participating institutions enter into agreements with the VA to contribute additional funding beyond the GI Bill's maximum tuition and fee coverage. This ensures that veterans can pursue their desired education without incurring excessive financial burdens. The Yellow Ribbon Program has been instrumental in expanding educational opportunities for veterans at a wide range of institutions across the country.

While on active duty, service members have access to Military Tuition Assistance (TA) programs. These programs provide financial assistance for off-duty education and training, covering tuition and fees for courses taken at accredited institutions. Through TA, service members can pursue higher education while still in the military, making progress toward degrees or certifications that will benefit them in their postmilitary careers. TA programs vary across branches of the military, but they all emphasize the importance of continuous learning and professional development for service members.

Transition Assistance Programs are designed to provide comprehensive support to service members as they prepare to transition out of the military. These programs include workshops, seminars, and counseling sessions that cover various aspects of transitioning to civilian life, including education and training opportunities. TAPs help veterans explore their educational options, understand the benefits available to them, and develop strategies for successful career transitions. These programs are often mandatory for service members nearing separation or retirement, ensuring that they are well-informed and prepared for their postmilitary journey.

The availability of education and training benefits for veterans is crucial in empowering them to pursue their educational and career aspirations after military service. Programs such as the Post-9/11 GI Bill, VR&E Program, Yellow Ribbon Program, Military Tuition Assistance, and Transition Assistance Programs provide the necessary financial support and resources for veterans to access higher education, vocational training, and skill development. By investing in veterans' education and training, societies honor their service and enable them to build fulfilling and successful postmilitary lives.

4. Mental Health Support: (a) Counseling and Therapy: Veterans transitioning out of the service often face mental health challenges, including PTSD, depression, and anxiety. The VA provides mental health support through counseling, therapy, and specialized programs tailored to the unique needs of veterans. These services aim to promote mental well-being, aid in the adjustment to civilian life, and prevent or manage mental health crises.

The mental health and well-being of veterans are of utmost importance, given the unique challenges they may face as a result of their military service. Recognizing this, the VA has established comprehensive mental health support programs to address the specific needs of veterans. The VA offers a wide range of mental health care services specifically tailored to meet the needs of veterans. These services encompass individual and group therapy, psychiatric evaluations, medication management, and crisis intervention. The VA's mental health care providers are equipped with specialized knowledge and experience in working with veterans, enabling them to understand and address the unique challenges faced by this population. The VA's commitment to evidence-based practices ensures that veterans receive the most effective and up-to-date treatments available.

Tragically, veterans are at an increased risk of suicide compared to the general population. Recognizing this alarming statistic, the VA has implemented a comprehensive suicide prevention program aimed at reducing the incidence of veteran suicide. This program includes mental health assessments, crisis-intervention services, outreach and education campaigns, and the Veterans Crisis Line—an invaluable resource available 24/7 for veterans in need of immediate assistance. The VA's suicide prevention efforts prioritize early intervention, fostering a culture of care, and providing the necessary support to prevent tragic outcomes.

Recognizing the need to provide accessible mental health care, particularly for veterans residing in remote areas or facing mobility challenges, the VA has expanded its telehealth services. Through telehealth, veterans can access mental health care and counseling remotely, using videoconferencing technology. This innovation has increased access to care, reduced travel burdens, and ensured that veterans receive the support they need regardless of their geographical location.

The VA's mental health support services are essential in addressing the specific challenges faced by veterans and promoting their overall well-being. Through comprehensive mental health care, suicide prevention initiatives, specialized PTSD programs, substance-abuse treatment, and the use of telehealth services, the VA ensures that veterans have access to the support they need. Society must recognize the sacrifices made by veterans and prioritize their mental health care, honoring their service by providing the necessary resources and support to lead fulfilling and healthy lives. By continuing to invest in mental health support for veterans, we demonstrate our commitment to their well-being and acknowledge their invaluable contributions to our nation.

The health care and benefits provided to veterans transitioning out of the service play a vital role in facilitating a successful and smooth transition into civilian life. Access to comprehensive health-care services, disability benefits, educational assistance, and mental health support not only addresses the specific health needs of veterans but also empowers them to build meaningful careers and lead fulfilling lives. By recognizing the sacrifices and service of veterans and ensuring their well-being, society can honor their commitment and provide a strong foundation for their future endeavors.

Health-care benefits play a crucial role in supporting veterans' overall well-being and quality of life. These benefits not only ensure access to essential medical services but also have a significant impact on veterans' housing opportunities. Access to comprehensive health-care benefits enhances veterans' ability to maintain stable housing. By receiving regular medical care, veterans can address health issues promptly, preventing the escalation of conditions that may lead to housing instability. Health-care benefits cover preventive services, routine checkups, and treatment for both physical and mental health conditions. This proactive approach to health care ensures that veterans can address health concerns before they become severe, reducing the likelihood of housing disruptions due to health-related emergencies.

Health-care benefits often include access to medications and specialized treatments. By providing veterans with necessary medical interventions, these benefits contribute to overall well-being, allowing veterans to remain healthy and active in their communities. When veterans can maintain their health, they are more likely to sustain stable housing situations, leading to increased housing security and improved overall quality of life.

Health-care benefits specifically aimed at addressing veterans' mental health needs have a direct impact on housing stability. Mental health conditions, such as PTSD or depression, can significantly impact veterans' housing situations, leading to homelessness or housing insecurity. By providing mental health services, including therapy, counseling, and medication management, health-care benefits enable veterans to effectively manage their mental health conditions.

When veterans have access to mental health support, they are better equipped to address trauma-related issues, manage stress, and develop coping strategies. Improved mental well-being translates into greater stability in housing situations, as veterans can maintain healthier relationships, secure employment, and navigate life transitions more effectively. Health-care benefits that address mental health concerns contribute to housing stability by addressing the underlying causes of housing insecurity and providing the necessary support for veterans to thrive in their housing environments.

Health-care benefits for veterans often extend beyond medical treatment to include supportive services that facilitate community integration and housing opportunities. Programs such as case management, vocational rehabilitation, and peer support services are integral to veterans' successful reintegration into civilian life and housing stability.

Through case management, veterans receive personalized support in navigating housing resources, securing stable housing, and accessing necessary social services. Vocational rehabilitation programs help veterans develop job skills, enhance employability, and increase income potential, which in turn strengthens their housing prospects. Peer support services create networks and communities of veterans who can provide guidance, understanding, and assistance in various aspects of life, including housing-related matters.

Health-care benefits provided to veterans include targeted programs aimed at preventing homelessness and assisting those experiencing housing crises. These programs often integrate health-care services with housing support, ensuring that veterans have access to stable housing options and the necessary resources to sustain their housing situations. By combining health-care benefits with housing assistance, veterans can access a comprehensive continuum of care that addresses both their health and housing needs.

These homelessness prevention programs may include rental assistance, transitional housing, and emergency shelter services. By combining these housing interventions with health-care benefits, veterans experiencing homelessness or housing instability can access the necessary support to regain housing stability and work toward long-term housing solutions.

Health-care benefits for veterans have a significant impact on housing opportunities by promoting stability, addressing mental health needs, facilitating community integration, and preventing homelessness. Access to comprehensive health-care ensures that veterans can maintain their health, address health issues proactively, and receive specialized mental health support.

Chapter 10

Housing

Ensuring housing stability and security is a vital component of supporting veterans as they transition to civilian life. Recognizing the unique challenges veterans may face, numerous housing resources and programs have been established to assist veterans in securing suitable and affordable housing. This chapter explores the range of housing resources available for veterans, including government initiatives, nonprofit organizations, and specialized programs aimed at addressing veterans' housing needs. These resources play a crucial role in empowering veterans with the stability and security they deserve in their postmilitary lives.

1. VA Housing Programs: The Department of Veterans Affairs offers various housing programs tailored specifically for veterans. These programs aim to provide affordable housing options, prevent homelessness, and support veterans in maintaining stable housing situations. The following are some notable VA housing programs:

The HUD-VASH program combines rental assistance from the Department of Housing and Urban Development (HUD) with case management and clinical services provided by the VA. This initiative targets veterans experiencing homelessness or at risk of becoming homeless, offering them housing vouchers and supportive services to help them secure and maintain housing stability.

Homelessness among veterans is a significant concern that demands comprehensive solutions. The HUD-VASH (Department of Housing and Urban Development - Veterans Affairs Supportive Housing) program is a collaborative effort between the VA and the HUD to address the housing needs of homeless veterans. The HUD-VASH program combines rental assistance from HUD with case management and clinical services provided by the VA. The program aims to provide long-term, stable housing solutions for homeless veterans by offering them housing vouchers and support services to address their specific needs. The collaboration between HUD and the VA ensures that both housing and comprehensive support are provided, recognizing the complex challenges veterans may face.

Eligible homeless veterans receive housing vouchers through the HUD-VASH program, which assists in subsidizing a portion of their rent for privately owned housing. These vouchers are administered by local public housing agencies (PHAs) and are based on a joint allocation between HUD and the VA. The rental assistance component of the program helps veterans access safe, affordable, and stable housing, laying the foundation for their successful reintegration into civilian life.

A distinguishing feature of the HUD-VASH program is the provision of comprehensive case management and support services. Each veteran participating in the program is assigned a VA case manager who assists in developing an individualized housing stability plan. Case managers provide ongoing support, connecting veterans with necessary resources and services, including mental health care, substance-abuse treatment, employment assistance, and access to health care.

The combination of rental assistance and case management allows veterans to address the underlying issues that contributed to their homelessness, empowering them to regain stability and self-sufficiency. The supportive services provided by the VA in partnership with community organizations play a vital role in helping veterans overcome barriers and maintain housing stability in the long term.

With a specific focus on homeless veterans, the HUD-VASH program directs its efforts toward those with the most pressing needs. Homeless veterans are identified through local coordinated entry systems, which prioritize individuals based on vulnerability and the length of time spent without housing. This targeted approach ensures that the program reaches those veterans who are most in need of housing assistance and support.

The HUD-VASH program's success is rooted in its collaborative approach, bringing together the expertise and resources of both the HUD and the VA. Close coordination between the two departments enables seamless integration of housing assistance and supportive services. PHAs, VA medical centers, local homeless service providers, and community-based organizations work together to identify eligible veterans, secure suitable housing options, and provide ongoing support. This collaborative effort maximizes the impact of the program and facilitates effective service delivery to veterans.

Positive outcomes have been observed through the implementation of the HUD-VASH program, effectively reducing homelessness among veterans and fostering stability in housing situations. According to the VA's Annual Homeless Assessment Report, the HUD-VASH program has contributed significantly to the overall decline in veteran homelessness since its inception. The program has been successful in increasing access to permanent housing, improving housing retention rates, and supporting veterans in their efforts to reintegrate into their communities.

By embodying a collaborative and comprehensive approach, the HUD-VASH program aims to combat veteran homelessness and advance housing stability. By combining rental assistance with comprehensive case management and support services, the program offers homeless veterans a pathway to stable and sustainable housing. Through the joint efforts of HUD and the VA, the HUD-VASH program continues to make a significant impact in reducing veteran homelessness and empowering veterans with the resources and support they need to secure the fundamentals of survival.

Specially Adapted Housing (SAH) and Special Home Adaption (SHA) grants provide financial assistance to eligible veterans with disabilities to help them modify or adapt their homes to meet their specific needs. These grants enable veterans to live independently and comfortably in their own homes, enhancing their quality of life and overall well-being.

The SAH program is a vital resource offered by the VA to support veterans with disabilities in achieving housing independence and accessibility. SAH grants provide financial assistance for eligible veterans to modify or adapt their homes to accommodate their unique needs.

The SAH program aims to provide housing solutions that enhance independence, mobility, and accessibility for veterans with service-connected disabilities. Through SAH grants, veterans are empowered to make necessary modifications or adaptations to their homes, enabling them to live more comfortably and independently. The program recognizes the importance of creating safe and accessible living environments that cater to the unique needs of disabled veterans.

To qualify for SAH grants, veterans must have a permanent and total service-connected disability related to their military service. Disabilities may include loss or loss of use of both legs, loss or loss of use of both arms, severe burn injuries, or blindness. The eligibility criteria ensure that SAH grants are targeted toward veterans with the most significant and life-altering disabilities.

SAH grants offer financial support to eligible veterans, enabling them to customize their homes according to their unique requirements. These grants can be utilized for a range of modifications and adaptations aimed at enhancing accessibility and independence. Examples include constructing or renovating existing homes to accommodate wheelchair access through the installation of ramps, widened doorways, and accessible bathrooms. Specialized equipment like stair lifts, elevators, or modified kitchen fixtures can also be installed to promote ease of movement within the home. Additionally, adaptations specific to disability-related needs, such as lowered countertops, grab bars, roll-in showers, or sensory modifications for individuals with visual or hearing impairments, are covered by these grants.

The SAH grants cover a significant portion of the costs associated with these modifications, alleviating the financial burden on veterans and their families. The grants enable veterans to create living spaces that are safe, functional, and tailored to their specific requirements, promoting a higher quality of life and greater independence.

The application process for SAH grants involves working closely with the VA and its regional offices. Veterans must submit the necessary documentation, including medical evidence of their service-connected disabilities and detailed plans for the proposed modifications. The VA evaluates each application on an individual basis, considering the veteran's specific needs and eligibility criteria.

The VA assists in the application process, including guidance on necessary documentation, coordination with contractors and builders, and oversight to ensure that the modifications meet applicable standards and regulations. The VA's commitment to supporting veterans throughout the process streamlines the grant application and implementation, making it more accessible for veterans with disabilities.

The SAH program has a profound impact on the lives of veterans with disabilities, enhancing their independence, mobility, and overall well-being. By providing the financial means to adapt to their homes, SAH grants empower veterans to age in place, maintain family unity, and remain active members of their communities. Accessible and modified housing environments not only promote physical independence but also contribute to veterans' mental and emotional well-being, reducing the daily challenges and barriers associated with disability.

The SAH program also supports veterans in reintegrating into their communities. By facilitating accessible housing, the program encourages social participation and community engagement, enabling veterans to interact with neighbors, participate in local activities, and lead fulfilling lives.

The VA's Home Loan Guaranty Program assists veterans in obtaining favorable home loan terms, including lower interest rates, with the VA guaranteeing a portion of the loan. This program aims to make homeownership more accessible and affordable for veterans, facilitating their transition to civilian life and promoting long-term stability.

The VA Home Loan Guaranty Program is a valuable resource provided by the VA to support veterans in achieving the dream of homeownership. This program guarantees loans made by private lenders to eligible veterans, enabling them to secure favorable loan terms and overcome traditional barriers to homeownership. The VA Home Loan Guaranty Program aims to help eligible veterans, service members, and surviving spouses obtain home loans with favorable terms. Instead of providing loans directly, the VA guarantees a portion of the loans made by private lenders, reducing the risk for lenders and making homeownership more accessible for veterans. The program recognizes the importance of stable housing and financial security in the postmilitary lives of veterans.

To qualify for the VA Home Loan Guaranty Program, veterans must meet certain eligibility requirements. These criteria may include the following:

- Veterans must have served a minimum period of active duty service, typically 90 consecutive days during wartime or 181 days during peacetime. Different requirements may apply for National Guard and Reserve members.
- Veterans must have been discharged under conditions other than dishonorable.
- Veterans need to obtain a Certificate of Eligibility (COE), which verifies their eligibility for the program. The COE can be obtained through the VA or requested through the lender.

The VA Home Loan Guaranty Program offers several significant benefits for veterans seeking to purchase or refinance a home:

One of the most significant advantages of the program is the ability to purchase a home without a down payment. This feature eliminates the need for substantial upfront cash, making homeownership more attainable for veterans, especially those with limited financial resources.

The VA guarantees a portion of the loan, reducing the risk for lenders and enabling them to offer more favorable loan terms to veterans. These terms may include competitive interest rates, flexible credit requirements, and no mortgage insurance requirements, resulting in lower monthly payments and long-term savings for veterans.

In the unfortunate event that a veteran faces financial hardship and struggles to make mortgage payments, the VA provides assistance and counseling to help veterans avoid foreclosure. The VA is committed to supporting veterans in maintaining homeownership and preventing housing instability.

Applying for a VA home loan involves working with a private lender that participates in the VA Home Loan Guaranty Program. Veterans need to provide the necessary documentation, including proof of income, credit history, and the COE obtained from the VA. Lenders guide veterans through the application process, ensuring they understand the requirements and benefits of the program.

The VA also provides assistance and resources to help veterans navigate the home-buying process. Housing counselors and loan specialists are available to address any questions or concerns veterans may have, ensuring they have access to the support they need throughout the loan application and homeownership journey.

The VA Home Loan Guaranty Program has a significant impact on veterans' housing stability and overall well-being. By eliminating the need for a down payment and offering favorable loan terms, the program makes homeownership a reality for many veterans who may have otherwise faced barriers to entry into the housing market. Stable homeownership provides veterans with a sense of security, pride, and belonging, allowing them to establish roots in their communities and build a foundation for their families' futures.

2. Nonprofit Organizations and Veterans Service Organizations: Numerous nonprofit organizations and veterans service organizations (VSOs) are dedicated to providing housing resources and support to veterans. These organizations often work in collaboration with the VA and other government agencies to address the specific housing needs of veterans.

The VFW offers financial grants through its Unmet Needs program to assist veterans in covering housing-related expenses, such as rent, mortgage payments, or utility bills, during times of financial hardship.

The Veterans of Foreign Wars (VFW) is a renowned veterans' service organization that plays a vital role in supporting transitioning service members as they navigate their postmilitary lives. With a rich history and a nationwide network of chapters, the VFW offers a range of programs and resources aimed at assisting veterans during the transition process.

The VFW provides transitioning service members with a strong support network of fellow veterans who have experienced similar challenges. Through local VFW chapters, veterans can connect with others who understand the unique transition from military to civilian life. This network serves as a valuable source of camaraderie, mentorship, and guidance, creating an environment where transitioning service members can share experiences, seek advice, and receive support from those who have walked the same path.

The VFW offers a range of transition assistance programs designed to facilitate a successful transition from military service to civilian life. These programs aim to address key areas of concern for transitioning service members, including employment, education, health care, and personal well-being. Some notable VFW programs include the following:

- The VFW assists transitioning service members with finding meaningful employment opportunities through job fairs, career counseling, resume assistance, and networking events. The organization also advocates for policies and initiatives that promote veteran employment and job retention.
- The VFW provides scholarship opportunities and educational resources to help transitioning service members pursue higher education or vocational training. These programs aim to facilitate a smooth transition into the civilian workforce and enhance veterans' career prospects.
- The VFW promotes physical and mental well-being through various health and wellness programs. These initiatives address issues such as post-traumatic stress disorder (PTSD), traumatic brain injury (TBI), and other challenges commonly faced by transitioning service members. The VFW connects veterans with resources, support networks, and counseling services to ensure their overall well-being.
- The VFW plays a crucial role in advocating for policies and legislation that benefit transitioning service members. Through its extensive advocacy efforts, the VFW ensures that the concerns and needs of transitioning veterans are heard at local, state, and national levels. The organization's collective voice helps shape policies related to employment, health care, education, and other areas of importance to transitioning service

members. The VFW's advocacy work serves as a powerful tool in creating a supportive environment for veterans during their transition and beyond.

Navigating the complex landscape of benefits and entitlements can be challenging for transitioning service members. The VFW provides guidance and assistance in understanding and accessing the various benefits available to veterans, including disability compensation, health-care services, and housing support. Through their knowledgeable service officers, the VFW helps transitioning service members navigate the paperwork and application processes, ensuring they receive the benefits they are entitled to.

The VFW encourages transitioning service members to engage with their local communities through volunteering and community service. This involvement not only provides veterans with a sense of purpose and fulfillment but also helps them build strong networks, develop new skills, and establish meaningful connections in their civilian lives. The VFW's emphasis on community engagement reinforces the idea of continued service and encourages transitioning service members to make positive contributions to their communities.

Operation Homefront provides various housing assistance programs, including rent-free transitional housing for veterans and their families, financial assistance for rent and utilities, and home-repair grants.

Operation Homefront is a renowned nonprofit organization that focuses on supporting transitioning veterans and their families as they navigate the challenges of returning to civilian life. With a strong commitment to honoring veterans and providing assistance in various areas, Operation Homefront offers programs and resources aimed at easing the transition process. Operation Homefront provides emergency financial assistance to transitioning veterans who may be facing financial hardships. This support is crucial during the transition process, as veterans adjust to new employment, housing, and financial responsibilities. The organization offers direct financial aid for essentials such as rent, utilities, groceries, and transportation, ensuring that veterans and their families can maintain stability during this critical time.

Operation Homefront understands the importance of secure housing for transitioning veterans. The organization offers transitional housing programs that provide temporary accommodations to veterans and their families while they seek permanent housing solutions. These programs bridge the gap between military service and civilian life, offering a safe and supportive environment where veterans can focus on finding stable, long-term housing.

Through its Homes on the Homefront program, Operation Homefront offers mortgage-free homes to transitioning veterans and their families. These homes are donated by various organizations and individuals who recognize the sacrifices made by veterans and their dedication to serving the country. This program not only provides housing stability but also relieves the financial burden of homeownership, allowing veterans to focus on their transition and future endeavors.

Operation Homefront assists transitioning veterans in their career transition by offering programs and resources aimed at enhancing their job prospects and professional development. These initiatives may include career counseling, job-placement assistance, resume building, interview preparation, and networking opportunities. By equipping veterans with the necessary tools and guidance, Operation Homefront empowers them to secure meaningful employment and establish successful civilian careers.

Recognizing the importance of family stability during the transition process, Operation Homefront offers various programs and resources to support veterans' families. This includes financial assistance for childcare expenses, back-to-school supplies, and holiday assistance. By easing the financial burden on veterans' families, Operation Homefront ensures that the entire family unit receives the support needed for a successful transition.

Operation Homefront actively encourages transitioning veterans to engage with their local communities through volunteer opportunities. By participating in community-service activities, veterans can build connections, develop new skills, and contribute to the well-being of their communities. This involvement reinforces the idea of continued service and helps veterans establish a sense of purpose and belonging in their civilian lives.

Transitioning from military to civilian life can be emotionally challenging for veterans. Operation Homefront provides resources and support to address the emotional well-being of transitioning veterans and their families. This may include counseling services, support groups, and referrals to mental health professionals. By fostering a supportive environment, Operation Homefront helps veterans and their families navigate the emotional complexities of the transition process.

Operation Homefront serves as a valuable resource for transitioning veterans, offering a range of programs and support to ease the challenges associated with returning to civilian life. By providing emergency financial assistance, transitional housing, career-transition support, family-support programs, and emotional guidance, Operation Homefront empowers veterans to successfully transition and thrive in their postmilitary lives. The organization's commitment to serving veterans and their families is a testament to the profound impact it has on their well-being and long-term success.

Habitat for Humanity's Veterans Build program engages veterans and volunteers to build and repair homes for low-income veterans and their families. This initiative promotes community involvement and provides affordable homeownership opportunities for veterans.

Habitat for Humanity Veterans Build is a unique program within the renowned nonprofit organization Habitat for Humanity that specifically focuses on supporting veterans and their families in achieving affordable homeownership. By harnessing the power of community engagement and volunteerism, Habitat for Humanity Veterans Build aims to address the housing needs of veterans and create a positive impact on their lives.

The Habitat for Humanity Veterans Build program is driven by a mission to honor veterans and their service by providing them with the opportunity to own safe, decent, and affordable homes. The program seeks to address the unique housing challenges faced by veterans and their families, recognizing the importance of stable housing as a foundation for successful postmilitary lives. The program's objectives include constructing or rehabilitating homes, fostering community engagement, and supporting veterans throughout the homeownership process.

A key aspect of the Habitat for Humanity Veterans Build program is the construction and rehabilitation of homes specifically designated for veterans and their families. Through the efforts of volunteers, community partners, and the veterans themselves, Habitat for Humanity builds or renovates homes to meet the specific needs of veterans, taking into consideration accessibility requirements and other factors. The program not only provides housing but also fosters a sense of pride and accomplishment for veterans who actively participate in the building process.

The Habitat for Humanity Veterans Build program emphasizes the power of community engagement and volunteerism in supporting veterans. Local communities come together to support the construction or renovation of homes, providing a platform for individuals, businesses, and organizations to actively contribute their time, skills, and resources. This engagement not only helps create affordable housing but also builds connections, fosters understanding, and bridges the gap between veterans and their communities.

As part of the Habitat for Humanity model, the Veterans Build program incorporates the concept of "sweat equity," whereby veterans actively participate in the construction or renovation of their own homes. By investing their time and effort, veterans contribute to the creation of their future homes while developing new skills and a sense of ownership. Additionally, Habitat for Humanity offers homeownership education and financial-literacy programs to equip veterans with the knowledge and tools necessary to successfully navigate the responsibilities of homeownership.

One of the primary benefits of the Habitat for Humanity Veterans Build program is the opportunity for veterans to achieve affordable homeownership. Through the program, veterans can obtain homes at an affordable price, typically with low or no-interest mortgages. This enables veterans to establish stability, build equity, and experience the pride and security of owning their own homes. The financial benefits extend beyond the immediate purchase, as affordable homeownership can have long-term positive impacts on veterans' financial well-being and overall quality of life.

Habitat for Humanity Veterans Build recognizes that successful homeownership goes beyond the physical structure of a home. The program provides additional supportive services to veterans and their families, including financial counseling, home-maintenance workshops, and access to community resources. This holistic approach aims to ensure that veterans have the necessary support to maintain their homes and integrate seamlessly into their communities, fostering long-term stability and well-being.

Habitat for Humanity Veterans Build is a powerful program that combines the vision of affordable homeownership with community engagement and volunteerism to support veterans and their families. By constructing or rehabilitating homes, fostering community connections, and providing comprehensive support throughout the homeownership process, Habitat for Humanity empowers veterans to rebuild their lives and build a foundation for their future.

3. State and Local Housing Initiatives: Many states and local municipalities offer housing resources and programs specifically designed for veterans. These initiatives vary by location but can include rental assistance programs, affordable housing developments reserved for veterans, and partnerships with local landlords to provide housing options. Veterans can access these resources through state and local housing agencies or veterans affairs offices.

State and local governments across the United States recognize the importance of supporting veterans in their transition to civilian life, including providing access to safe and affordable housing. Various housing initiatives have been implemented at the state and local levels to address the unique housing needs of veterans and ensure they have a stable foundation as they reintegrate into society. State and local housing initiatives for veterans share common objectives aimed at addressing the housing challenges faced by this specific population. These initiatives seek to provide affordable housing options, prevent homelessness, enhance accessibility, and foster community integration for veterans. The programs are designed to recognize the sacrifices made by veterans and ensure they have access to safe and stable housing, promoting their overall well-being and successful transition into civilian life.

Many states and local governments offer homeownership assistance programs tailored specifically for veterans. These programs provide financial assistance, including down payment and closing-cost assistance, to help veterans purchase homes. They may also offer favorable mortgage terms and interest rates, making homeownership more accessible and affordable. Homeownership assistance programs empower veterans to establish roots in their communities, build equity, and create a stable foundation for their families.

To address the issue of housing affordability, state and local governments often provide rental-assistance programs and housing vouchers for veterans. These initiatives help offset the cost of rent and ensure that veterans have access to safe and affordable housing options. Rental assistance programs may include subsidies, rental vouchers, or supportive housing arrangements, allowing veterans to secure stable housing while managing their financial resources during the transition process.

Recognizing the unique needs of veterans with disabilities or mobility limitations, many state and local housing initiatives offer programs that focus on home modification and accessibility. These programs provide financial assistance or grants to modify existing homes or build new homes to accommodate specific accessibility requirements. Modifications may include wheelchair ramps, widened doorways, bathroom modifications, and other adaptations that enable veterans with disabilities to live independently and comfortably.

Addressing homelessness among veterans is a priority for state and local housing initiatives. These programs offer supportive services and resources to help veterans at risk of homelessness or those already experiencing homelessness. Supportive services may include case management, counseling, job placement assistance, substance-abuse treatment, and mental health support. Through comprehensive approaches that combine housing assistance with support services, state, and local initiatives strive to prevent and reduce homelessness among veterans.

State and local governments often collaborate with nonprofit organizations specializing in veterans' housing to maximize their impact. These partnerships leverage the expertise and resources of nonprofit organizations to implement targeted housing programs and provide additional support to veterans. Nonprofit organizations may offer transitional housing, emergency shelter, rental assistance, and supportive services, further enhancing the housing options and assistance available to veterans at the state and local levels.

State and local housing initiatives for veterans emphasize the importance of community integration and support. These programs encourage collaboration between veterans, local communities, and service providers to foster a sense of belonging and connection. Community-based initiatives may include veteran-specific housing developments, community events, mentorship programs, and employment assistance, all aimed at helping veterans reintegrate into civilian life and build strong support networks.

4. Supportive Services for Veteran Families (SSVF): The Supportive Services for Veteran Families program, funded by the VA, provides grants to nonprofit organizations and VSOs to offer a range of housing-related services to low-income veterans and their families. These services include case management, financial assistance for rent and utilities, and supportive counseling to help veterans maintain stable housing.

Supportive Services for Veteran Families (SSVF) is a vital program that provides a comprehensive range of support services to veterans and their families who are experiencing or at risk of homelessness. Administered by the VA, SSVF aims to address the unique needs of veterans and ensure they have access to stable housing, financial assistance, and supportive resources. The primary objective of the SSVF program is to prevent and end homelessness among veterans and their families. The program focuses on three core areas: housing stability, financial assistance, and supportive services. By targeting these key components, SSVF aims to address the root causes of homelessness and provide veterans with the necessary resources and support to regain stability and thrive.

One of the fundamental aspects of the SSVF program is ensuring housing stability for veterans and their families. SSVF provides various services to achieve this, including housing counseling, assistance with locating affordable housing options, and help with obtaining or maintaining a lease. The program also offers case management to address any barriers to housing stability and works closely with landlords to secure stable housing arrangements for veterans.

SSVF recognizes that financial challenges can often contribute to homelessness or housing instability. To address this, the program provides financial assistance to eligible veterans and their families. This may include rental and utility deposit assistance, rental and utility payment support, and assistance with moving costs. By alleviating the financial burden, SSVF helps veterans secure and maintain stable housing situations.

SSVF offers a wide range of supportive services tailored to the unique needs of veterans and their families. These services encompass comprehensive case management, employment assistance, access to health-care services, mental health counseling, substance-abuse counseling, and assistance with accessing benefits and entitlements. By addressing the holistic needs of veterans, SSVF helps them navigate various challenges and establish a strong foundation for long-term stability.

A significant component of the SSVF program is rapid rehousing, which focuses on quickly transitioning homeless veterans and their families into permanent housing. Rapid rehousing interventions include short-term rental assistance, case management, and other supportive services aimed at helping veterans secure and maintain stable housing situations. This approach prioritizes swift action to prevent prolonged periods of homelessness and supports the transition to self-sufficiency.

SSVF operates through partnerships with community-based organizations and nonprofit agencies. These organizations play a critical role in delivering services on the ground and facilitating the implementation of the program. SSVF collaborates with community partners to expand housing options, establish connections with local service providers, and enhance resources available to veterans and their families. These partnerships enable a more coordinated and effective approach to addressing veteran homelessness at the local level.

The SSVF program has demonstrated significant success in preventing and ending homelessness among veterans. Through its comprehensive support services, SSVF has helped countless veterans and their families regain stability, secure permanent housing, and rebuild their lives. The program's focus on housing stability, financial assistance, and supportive services ensures a holistic approach to addressing the complex needs of veterans experiencing homelessness or housing instability.

SSVF plays a vital role in the prevention and resolution of homelessness among veterans and their families. Through its focus on housing stability, financial assistance, and supportive services, SSVF empowers veterans to overcome homelessness, regain stability, and thrive in their transition."

Veterans, as esteemed members of the military community, deserve the utmost support and care as they transition back into civilian life. One critical aspect of this transition is ensuring access to stable and affordable housing. Throughout this chapter, we have explored the numerous housing opportunities and initiatives available to veterans, highlighting the significant impact they have on improving the lives of those who have served their country.

From federal programs such as the VA Home Loan Guaranty Program, which provides veterans with favorable mortgage terms, to state and local initiatives like rental-assistance programs and accessibility modifications, the range of housing opportunities for veterans is diverse and robust. These programs aim to address the unique challenges faced by veterans and their families, including housing affordability, homelessness prevention, and supportive services.

One common thread running through these housing opportunities is the recognition of veterans' sacrifices and the desire to provide them with stability, security, and a sense of community. Housing initiatives not only offer physical shelter but also foster a sense of belonging and a supportive environment where veterans can thrive. They prioritize the holistic well-being of veterans, acknowledging that a stable home is a foundation for success in other areas of life.

These housing opportunities go beyond mere shelter. They provide educational and training benefits, mental health support, and employment assistance, all of which contribute to the overall well-being and successful transition of veterans. By combining housing with comprehensive support services, veterans are equipped with the tools they need to overcome challenges, rebuild their lives, and integrate into their communities.

Community engagement and partnerships are essential components of many housing programs for veterans. Local communities, nonprofit organizations, and governmental agencies join forces to ensure that veterans receive the necessary support and resources. This collaboration strengthens the fabric of society, fostering connections between veterans and their communities and promoting a sense of gratitude and unity.

While significant strides have been made in improving housing opportunities for veterans, it is important to acknowledge that challenges persist. Veterans transitioning out of the military may face difficulties such as a lack of affordable housing options, limited job prospects, or difficulties in navigating the complex process of accessing benefits. Continued advocacy, innovation, and collaboration are necessary to address these challenges and ensure that every veteran has access to safe and affordable housing.

Providing housing opportunities for veterans is not only a moral imperative but also a practical investment in the well-being of those who have served our country. By supporting veterans in securing stable and affordable housing, we honor their service, recognize their sacrifices, and contribute to their successful transition into civilian life. The combination of housing options, supportive services, and community engagement creates a solid foundation from which veterans can rebuild their lives, pursue their goals, and continue to make valuable contributions to society. As a grateful nation, we must ensure that veterans have every opportunity to thrive and find comfort and security in their postmilitary lives.

CONCLUSION

1. Identity and Purpose

The exploration of the identity and purpose of veterans reveals the profound impact of military service on individuals and the subsequent challenges they face when transitioning to civilian life. Throughout this journey, veterans undergo a transformative process as they navigate the complex interplay between their military identity and their search for renewed purpose in the civilian realm.

The identity of veterans is intricately linked to their military service, forged through shared experiences, values, and a strong sense of camaraderie. The military identity encompasses traits such as discipline, resilience, and a deep commitment to duty. However, as veterans transition to civilian life, they may grapple with reconciling their military identity with the demands and expectations of the civilian world. This struggle can sometimes lead to a loss of identity or a sense of disconnection, requiring introspection and adaptation to establish a new sense of self.

Finding purpose is a crucial aspect of the veteran experience. The military often provides a clear sense of purpose, as individuals serve a greater cause and work toward a shared mission. However, when veterans transition to civilian life, they may encounter challenges in defining and pursuing a new purpose. The search for the purpose involves introspection, self-discovery, and aligning personal values with meaningful endeavors. It may involve exploring new career paths, engaging in community service, or pursuing educational opportunities that resonate with veterans' passions and aspirations.

Supporting veterans in their journey of identity and purpose is crucial. Communities, employers, and policymakers can play a pivotal role in recognizing and valuing the unique skills, experiences, and perspectives that veterans bring. Building a bridge between military and civilian cultures, fostering understanding, and providing resources for personal and professional development can empower veterans to navigate their transition with confidence and purpose.

It is essential to create environments that encourage open dialogue and support networks, enabling veterans to connect with fellow veterans and civilians who can provide guidance, mentorship, and a sense of belonging. Programs and initiatives that promote the integration of veterans into various aspects of society, such as employment, education, and community engagement, contribute to a holistic approach to supporting their identity and purpose.

The identity and purpose of veterans are deeply intertwined with their military service and play a significant role in their successful transition to civilian life. By acknowledging the complexities of their identities and facilitating the pursuit of meaningful purpose, we can honor the sacrifices veterans have made and foster an inclusive and supportive society that enables them to thrive in their postmilitary lives. Through collective efforts, we can ensure that veterans find fulfillment, contribute their talents to society, and lead fulfilling and purposeful lives beyond their military service.

2. Civilian Culture Gaps

The examination of civilian culture gaps experienced by veterans sheds light on the unique challenges they face when transitioning from the military to civilian life. These gaps emerge due to fundamental differences in values, norms, and expectations between military and civilian cultures. Understanding and bridging these gaps is crucial for promoting successful integration and fostering a supportive environment for veterans.

The military culture instills a distinct set of values and behaviors, including discipline, hierarchy, and a strong sense of mission and duty. In contrast, civilian culture often prioritizes individualism, autonomy, and diverse perspectives. These differences can create a sense of disorientation and alienation for veterans as they navigate the civilian world.

One of the key civilian culture gaps is the disconnect in communication styles and language. The military employs a unique set of jargon, acronyms, and hierarchical structures that may be unfamiliar to civilians. This can hinder effective communication and understanding, leading to misinterpretation and frustration.

Another significant challenge is the perception and understanding of military experiences by civilians. Veterans may struggle to convey the complexity of their service and the emotional toll it can take, leading to a lack of empathy and comprehension from those who have not served. This disconnect can create feelings of isolation and make it difficult for veterans to relate to their civilian counterparts.

The transition from a highly structured and regimented military environment to the more flexible and decentralized nature of civilian life can be overwhelming for veterans. Adjusting to the absence of clear chains of command, established routines, and defined roles requires a period of adaptation and reorientation.

To address civilian culture gaps, awareness, education, and dialogue are crucial. Civilian communities can foster a greater understanding of the military experience through initiatives such as workshops, cultural exchange programs, and public events that facilitate interactions between veterans and civilians. These platforms provide an opportunity for veterans to share their experiences, challenge stereotypes, and foster mutual respect and understanding.

Employers play a vital role in bridging civilian culture gaps by recognizing and valuing the unique skills and experiences veterans bring to the workforce. Offering targeted training programs, mentorship, and creating a supportive work environment can facilitate the successful integration of veterans and help bridge the gap between military and civilian cultures.

Addressing civilian culture gaps is essential for supporting veterans in their transition to civilian life. By fostering understanding, empathy, and cultural exchange, we can create a more inclusive and supportive environment that recognizes and appreciates the contributions and sacrifices of veterans. Through collective efforts, we can bridge the gaps between military and civilian cultures, ensuring a smoother transition and promoting the well-being and success of our veterans in civilian life.

3. Translating Military Skills

Translating the military skills of veterans is a vital undertaking that holds significant importance in fostering successful postmilitary transitions. The unique skills, experiences, and attributes acquired during military service can be highly valuable and transferable to various civilian sectors. However, the process of translating these skills requires careful consideration and active collaboration between veterans, employers, educators, and policymakers.

Military veterans possess a wide range of skills that can be effectively utilized in civilian careers. These skills include leadership, discipline, teamwork, problem-solving, adaptability, resilience, and a strong work ethic. Veterans often possess technical expertise in fields such as engineering, logistics, communications, healthcare, and information technology. By recognizing and properly translating these skills, veterans can make substantial contributions to industries that require these competencies.

One of the primary challenges in translating military skills lies in overcoming the language barrier between military and civilian jargon. Veterans often use terminology and acronyms that are unfamiliar to civilians, making it difficult for employers to understand the depth and relevance of their experiences. It is essential to bridge this gap by creating standardized methods for translating military experience into language that is easily understood by civilian employers, such as through the development of military skills translators or certification programs.

Effective collaboration between veterans, employers, educators, and policymakers is crucial in creating an environment that supports and facilitates the translation of military skills. Employers should actively seek to understand and value the unique experiences and capabilities of veterans and adapt their recruitment processes to better assess and recognize these skills. Educators and training institutions can play a significant role by offering programs and resources that help veterans acquire additional civilian credentials and bridge any knowledge gaps.

Policymakers can also contribute by implementing supportive policies, such as tax incentives for employers hiring veterans or creating programs that facilitate skills translation and transition assistance.

Ultimately, translating the military skills of veterans is not only a responsibility but also an opportunity. It allows us to harness the immense potential and talent that veterans bring to the civilian workforce, leading to a more inclusive, diverse, and productive society. By recognizing and leveraging the skills gained during military service, we can help veterans successfully transition into meaningful and fulfilling careers, while also benefiting from their invaluable contributions to our communities and economy.

Translating military skills requires collaboration, understanding, and proactive efforts from multiple stakeholders. It is a worthwhile endeavor that can empower veterans and unlock their full potential in civilian life. By recognizing the value of military experience, bridging the language gap, and creating a supportive environment, we can honor the sacrifices and service of veterans while harnessing their immense talent for the betterment of our society.

4. Limited Civilian Networks

The limited civilian networks of veterans present a significant challenge that can hinder their successful transition into civilian life. After leaving the military, veterans often find themselves disconnected from the extensive support networks they had within the military community. This lack of civilian networks can impact various aspects of their lives, including employment opportunities, social integration, and access to resources and support systems.

The transition from military to civilian life can be a daunting process, and having a strong network of support is crucial. In the military, veterans were surrounded by peers who understood their experiences and shared a common bond. These networks provided a sense of camaraderie, mentorship, and access to information and resources. However, once veterans enter the civilian world, they may struggle to find a similar level of support and understanding.

Limited civilian networks can particularly affect veterans' employment prospects. Networking is a critical aspect of job search and career advancement, and without a robust civilian network, veterans may face difficulties in accessing job opportunities, receiving referrals, and obtaining necessary information about the civilian job market. This lack of connections can lead to increased challenges in securing meaningful employment that aligns with their skills and experiences.

The limited civilian networks of veterans can also impact their social integration and overall well-being. Feeling isolated or disconnected from the broader community can contribute to feelings of loneliness, depression, and anxiety. Veterans may find it challenging to build new friendships and establish social connections outside of the military context. This can further exacerbate the difficulties they face in their transition to civilian life.

Addressing the issue of limited civilian networks requires a multifaceted approach involving various stakeholders. Employers can play a vital role by actively seeking to hire veterans, creating veteran affinity groups within their organizations, and providing mentorship and networking opportunities. Community organizations, veteran service organizations, and nonprofit groups can organize events and programs that facilitate networking and social integration for veterans. Additionally, educational institutions and career counseling centers can offer resources and workshops on networking skills and techniques specifically tailored to veterans.

Technology can also serve as a valuable tool in bridging the gap of limited civilian networks. Online platforms and social media can provide virtual communities and support networks for veterans, connecting them with peers, mentors, and potential employers. These platforms can facilitate information sharing, professional development, and social interaction, offering a sense of belonging and support even in the absence of physical proximity.

Addressing the issue of limited civilian networks is crucial in supporting the successful transition of veterans into civilian life. By recognizing the challenges veterans face in establishing civilian networks and taking proactive measures to bridge the gap, we can provide the support and opportunities necessary for their social integration, employment success, and overall well-being. Building robust civilian networks for veterans not only benefits individuals but also contributes to the diversity, strength, and inclusivity of our communities as a whole.

5. Mental Health and Well-Being

The mental health and well-being of veterans is a critical concern that requires ongoing attention and support. Military service can have a profound impact on the psychological well-being of veterans, and we must recognize and address the unique challenges they face in maintaining good mental health.

The experiences of war, exposure to traumatic events, and the stress of military life can contribute to a range of mental health issues among veterans. Conditions such as PTSD, depression, anxiety, and substance abuse are commonly observed in this population. These conditions can significantly impact veterans' daily functioning, relationships, and overall quality of life.

It is crucial to prioritize the mental health needs of veterans and ensure they have access to appropriate care and support. Mental health services tailored to the unique experiences and needs of veterans should be readily available and easily accessible. This includes specialized programs and treatment approaches that take into account the specific challenges associated with military services, such as trauma-focused therapies, peer support groups, and employment and education assistance.

Additionally, destigmatizing mental health issues within the military and veteran community is essential. Encouraging open discussions about mental health and providing education and awareness campaigns can help reduce the barriers to seeking help and promote a culture of support and understanding. This involves challenging stereotypes and misconceptions surrounding mental health and emphasizing that seeking treatment is a sign of strength rather than weakness.

Collaboration between various stakeholders is crucial in addressing the mental health needs of veterans. This includes government agencies, health-care providers, community organizations, and veteran support networks. By working together, these stakeholders can ensure that veterans receive comprehensive and integrated care that addresses their mental health needs holistically.

Preventive measures are also vital in promoting the mental health and well-being of veterans. This includes early intervention, screening, and education on coping strategies and resilience-building. Efforts should be made to identify and address mental health concerns before they escalate and to equip veterans with the tools and resources they need to maintain their well-being throughout their lives.

Supporting the mental health and well-being of veterans is a moral imperative and a responsibility that society must bear. By providing accessible and specialized mental health services, combating stigma, and promoting a culture of support, we can help veterans lead fulfilling and productive lives beyond their military service. We must honor their sacrifices by ensuring they receive the care and support they need to thrive and enjoy a healthy and meaningful postservice life.

6. Financial Stability

The financial stability of veterans is a critical issue that requires attention and support from both the government and society as a whole. Despite their sacrifices and contributions to the nation, many veterans face unique financial challenges that can impede their ability to lead stable and fulfilling lives. Concerted efforts must be made to address these challenges and ensure the economic well-being of those who have served.

One of the key factors contributing to the financial instability of veterans is the transition from military to civilian life. This transition can be daunting, as veterans often face difficulties in finding employment that matches their skills and experiences. The skills acquired in the military, while valuable, may not always directly translate to the requirements of civilian job markets. Furthermore, issues such as physical or mental health disabilities resulting from military service can further complicate the employment prospects of veterans. Therefore, it is crucial to provide targeted job training, career counseling, and employment support to facilitate their successful integration into the civilian workforce.

Another significant aspect impacting veterans' financial stability is the availability and accessibility of comprehensive health care. Many veterans experience physical and mental health challenges, including PTSD, traumatic brain injuries, and other service-related disabilities. These conditions not only affect their overall well-being but also have economic consequences. Adequate and timely health care, including mental health services, is essential for veterans to recover and thrive. Ensuring that veterans have access to affordable and comprehensive health care, along with robust support systems, can alleviate the financial burdens associated with medical expenses.

The socioeconomic factors that veterans face must be considered. Homelessness, poverty, and substance abuse are unfortunate realities for some veterans. These issues often stem from a combination of factors such as limited job opportunities, insufficient financial resources, and mental health challenges. Addressing these underlying factors requires a multifaceted approach that includes affordable housing programs, targeted social services, and initiatives to combat substance abuse. By providing stable housing and support systems, veterans can have a solid foundation to rebuild their lives and improve their financial circumstances.

The government plays a crucial role in addressing the financial stability of veterans. Policies and programs, such as education and vocational training benefits, disability compensation, and pension plans, are essential components of the support system for veterans. It is vital for the government to continuously assess and improve these initiatives to ensure they meet the evolving needs of veterans.

However, it is not solely the responsibility of the government to support veterans financially. Civil society, private organizations, and individuals can also contribute to creating a supportive environment for veterans. Philanthropic efforts, community initiatives, and corporate partnerships can provide additional resources, mentorship programs, and employment opportunities. Society as a whole must recognize the sacrifices made by veterans and actively work toward creating an inclusive and supportive environment for them.

Addressing the financial stability of veterans requires a comprehensive and collaborative approach. By providing targeted job training, accessible health care, and addressing socioeconomic factors, we can support veterans in achieving stability and economic well-being. The government, society, and individuals need to recognize the challenges faced by veterans and actively contribute toward their financial security. Only by working together can we honor their service and ensure that they can lead prosperous and fulfilling lives after their military careers.

7. Education and Training

The education and training of veterans play a pivotal role in their successful transition from military to civilian life. By equipping veterans with the necessary knowledge, skills, and resources, we can empower them to pursue rewarding educational opportunities and secure meaningful employment. It is essential to recognize the unique experiences and qualifications of veterans and provide tailored support to facilitate their educational and training pursuits.

One of the key aspects of supporting veterans in education and training is the recognition of their prior military experience. The skills acquired during military service, such as leadership, discipline, problem-solving, and teamwork, are highly valuable in various academic and professional settings. Educational institutions and employers should acknowledge and credit these skills when evaluating veterans' qualifications. This recognition can enhance veterans' confidence, provide a smoother transition into civilian life, and open doors to diverse educational and training opportunities.

Veterans often face specific challenges when pursuing education and training, such as the need for flexible learning options and financial assistance. Many veterans are older than traditional college-age students, with family responsibilities and potentially limited financial resources. Flexible learning options, such as online or evening classes, can accommodate their unique circumstances and enable them to balance their educational pursuits with other commitments. Additionally, financial support in the form of scholarships, grants, and tuition assistance programs can alleviate the financial burden of education and training, making them more accessible to veterans.

To ensure the success of veterans in education and training, it is crucial to provide comprehensive support services. These services may include academic advising, counseling, and mentoring programs specifically tailored to the needs of veterans. Academic advisors can guide veterans in selecting suitable programs and courses that align with their career goals and interests. Counseling services can address the psychological and emotional challenges that veterans may face during the transition to civilian education. Mentorship programs can connect veterans with experienced professionals who can provide guidance, networking opportunities, and industry-specific insights.

Collaboration between educational institutions, government agencies, and private organizations is paramount in supporting veterans' education and training. Partnerships can facilitate the sharing of resources, expertise, and best practices, leading to more effective programs and services. By leveraging the collective efforts of various stakeholders, we can maximize the impact of educational and training initiatives for veterans.

The education and training of veterans are vital components of their successful integration into civilian life. By recognizing and valuing their military experience and providing flexible learning options, financial assistance, and comprehensive support services, we can empower veterans to pursue their educational and career aspirations. Collaboration between educational institutions, government agencies, and private organizations is crucial in ensuring the effectiveness and sustainability of these initiatives. Through these collective efforts, we can honor the commitment and sacrifices of veterans by equipping them with the tools and opportunities they need to thrive in their postmilitary lives.

8. Employment Opportunities

Employment opportunities for veterans are of paramount importance in ensuring their successful transition from military service to civilian life. Veterans possess unique skills, experiences, and qualities that make them valuable assets to the workforce. However, they often encounter challenges in finding suitable employment that fully utilizes their capabilities. It is crucial to address these challenges and provide robust support systems to facilitate the integration of veterans into the civilian job market.

One of the key factors impacting veterans' employment opportunities is the translation of their military skills to the civilian context. While veterans possess a wide range of valuable skills, such as leadership, teamwork, problem-solving, and adaptability, it is essential to bridge the gap between military and civilian terminology. Employers may not always fully understand or recognize the relevance of veterans' skills to their organizations. Therefore, initiatives such as skills-translation programs, resume-writing assistance, and interview preparation specifically tailored to veterans can enhance their marketability and increase their chances of securing suitable employment.

Veterans may face barriers related to the lack of professional networks and industry connections. Networking plays a crucial role in accessing job opportunities, and veterans may not have the same networks as their civilian counterparts. Providing mentorship programs, networking events, and career fairs that connect veterans with employers, industry professionals, and other veterans can help expand their professional networks. Establishing partnerships between employers and veteran-focused organizations can create pathways for veterans to enter various industries and secure employment.

Another significant aspect to consider is the provision of targeted training and educational opportunities to enhance veterans' employability. Skills gaps and technological advancements in the civilian workforce may require veterans to acquire additional training or education to compete effectively in the job market. Offering specialized training programs, vocational courses, and educational benefits can equip veterans with the necessary qualifications and certifications for in-demand industries. By aligning these programs with market needs and emerging sectors, veterans can access employment opportunities that match their skills and interests.

It is crucial to address potential biases and misconceptions surrounding veterans' employment. Some employers may hold misconceptions about veterans, such as concerns about PTSD or assumptions about rigid military hierarchies. Educating employers about the strengths, capabilities, and contributions of veterans can help dispel these misconceptions and foster a more inclusive and supportive work environment. Employer incentives, such as tax credits or recognition programs, can also encourage businesses to actively recruit and hire veterans.

Providing meaningful employment opportunities for veterans is essential to their successful reintegration into civilian life. By addressing the skills translation gap, facilitating networking and mentorship, offering targeted training and educational programs, and combating biases, we can unlock the immense potential of veterans in the workforce. Collaboration between government agencies, private-sector employers, veteran support organizations, and educational institutions is crucial in creating an environment that recognizes and values the unique contributions of veterans. Through these collective efforts, we can honor the service and sacrifices of veterans by ensuring their economic stability, professional growth, and long-term career success.

9. Health Care and Benefits

The health care and benefits provided to veterans are essential components of honoring their service and ensuring their well-being after military service. Veterans often face unique health-care needs resulting from their service-related physical and mental health challenges. It is crucial to prioritize their access to comprehensive health-care services and benefits to address these needs effectively.

One of the key aspects of supporting veterans' health care is the provision of accessible and high-quality medical services. Many veterans experience physical injuries, disabilities, and chronic health conditions resulting from their military service. Additionally, mental health conditions such as PTSD, depression, and anxiety are prevalent among veterans. Ensuring that veterans have access to timely and effective health care, including specialized treatment for service-related conditions, is crucial in promoting their overall well-being.

In addition to medical services, benefits such as disability compensation and pension plans are vital for veterans' financial stability. Disabilities acquired during military service can impact veterans' ability to work and lead fulfilling lives. Disability compensation provides financial support to veterans who have sustained service-related injuries or disabilities, helping them cope with the financial burden of medical care and daily living expenses. Pension plans, on the other hand, provide income support for veterans who have served for a certain period and have met specific eligibility criteria. These benefits are crucial in recognizing the sacrifices made by veterans and supporting their long-term financial security.

Another crucial aspect of health care and benefits for veterans is the provision of mental health services. The psychological toll of military service, including exposure to trauma and high-stress environments, can have long-lasting effects on veterans' mental well-being. It is imperative to ensure that mental health services, including counseling, therapy, and support programs, are readily available to veterans. These services can assist veterans in coping with the psychological challenges they may face and facilitate their successful reintegration into civilian life.

The coordination and integration of health-care services across different systems are vital for veterans. The VA plays a central rôle in providing health care and benefits to veterans. However, seamless coordination and collaboration between the VA, private health-care providers, and community organizations are essential to ensure that veterans receive timely and comprehensive care. Partnerships between these entities can help expand access to specialized care, reduce wait times, and improve the overall quality of health-care services for veterans.

Providing comprehensive health care and benefits to veterans is crucial for recognizing their sacrifices and ensuring their well-being after military service. Accessible and high-quality medical services, disability compensation, pension plans, and mental health support are all vital components of a robust support system for veterans. Collaboration between government agencies, health-care providers, community organizations, and private-sector stakeholders is essential in delivering integrated and effective health-care services to veterans. By honoring their service and addressing their unique health-care needs, we can repay the debt of gratitude owed to veterans and support them in leading healthy, fulfilling lives.

10. Housing

Housing for veterans is a crucial issue that demands immediate attention and effective solutions. Throughout history, veterans have selflessly served their countries and defended the principles of freedom and justice. However, upon returning home, many veterans face significant challenges in finding suitable housing, which can have detrimental effects on their overall well-being and successful reintegration into civilian life.

There is a pressing need for increased support and resources dedicated to addressing the housing needs of veterans. The scarcity of affordable and accessible housing options, combined with the unique challenges faced by veterans, such as physical and mental health issues, homelessness, and unemployment, requires a comprehensive approach that recognizes the complex nature of the problem.

To effectively tackle the issue of housing for veterans, collaboration among government agencies, nonprofit organizations, and the private sector is essential. Government entities should prioritize funding and policy initiatives aimed at expanding affordable housing options specifically tailored to veterans' needs. This can involve providing financial assistance, offering rental subsidies, and partnering with housing developers to create affordable and sustainable housing units.

Nonprofit organizations play a crucial role in providing direct support and services to veterans, including transitional housing, counseling, and job-placement programs. Their efforts should be further bolstered through increased funding and partnerships with other stakeholders. Additionally, private-sector involvement, such as corporate sponsorships, can help provide resources and expertise to address the unique challenges faced by veterans in securing housing.

It is imperative to prioritize early intervention and prevention strategies. Efforts should focus not only on assisting homeless veterans but also on preventing homelessness in the first place. This can be achieved by improving access to mental health services, employment opportunities, and supportive resources that address the underlying factors contributing to housing instability.

Housing for veterans is a matter of utmost importance that requires a multifaceted and collaborative approach. By combining government initiatives, nonprofit support, and private- sector involvement, we can make significant progress in addressing the housing needs of veterans. It is our moral duty to honor the sacrifices made by veterans by ensuring they have a safe and stable place to call home. Through concerted efforts, we can create a society that fully supports and appreciates those who have served our nation, providing them with the housing they deserve as they reintegrate into civilian life.

In conclusion, the successful transition of service members from military to civilian life is a critical endeavor that requires dedicated support, comprehensive resources, and collaborative efforts from various stakeholders. As service members complete their military service, they face unique challenges and opportunities as they navigate the transition process.

The transition from military to civilian life can be a complex and overwhelming experience for many service members. It involves not only finding suitable employment but also adapting to a new social and cultural environment, managing financial changes, and addressing physical and mental health needs. Recognizing these challenges, it is crucial to establish a robust support system that ensures a smooth and successful transition for our veterans.

One key aspect of supporting transitioning service members is the provision of comprehensive resources and guidance. Government agencies, such as the Department of Defense and the Department of Veterans Affairs, should collaborate to offer tailored programs that address the specific needs of transitioning service members. These programs should focus on career counseling, skills translation, job-placement assistance, and educational opportunities to help service members leverage their military experience in the civilian workforce.

Partnerships between the public and private sectors are vital in facilitating the transition process. Private employers should be encouraged to recognize the value and transferable skills that service members bring to the civilian workplace. Initiatives like mentorship programs, internships, and apprenticeships can help bridge the gap between military and civilian employees and provide valuable networking opportunities for transitioning service members.

In addition to employment support, it is crucial to prioritize the mental and physical health needs of transitioning service members. Accessible and effective health-care services, including mental health counseling, should be readily available to address the challenges of posttraumatic stress, depression, and other related conditions. Community-based organizations and nonprofit groups can also play a significant role in providing support networks, peer mentoring, and transitional housing for those in need.

Lastly, it is essential to foster a societal understanding and appreciation for the sacrifices made by service members. Raising awareness about the unique challenges faced during the transition process can help reduce stigmas and promote a more inclusive and supportive environment. Public campaigns, community outreach initiatives, and educational programs can contribute to creating a culture that values and understands the experiences of transitioning service members.

The successful transition of service members requires a collective effort and commitment from various stakeholders. By providing comprehensive resources, employment support, and health-care services, and fostering a supportive societal environment, we can ensure that our transitioning service members are equipped to thrive in civilian life. We must honor their service by facilitating a seamless transition and helping them build fulfilling and successful postmilitary careers.

RESOURCE LIST

1. Transition Assistance Programs (TAPs): Find local TAP offices and resources at the Department of Labor's official website: www.dol.gov/vets/programs/tap.
2. Department of Veterans Affairs (VA): Access a wide range of veteran services and benefits at the official VA website: www.va.gov.
3. Veterans Employment Center: Explore job opportunities, career tools, and resources for veterans at the Veterans Employment Center website: www.ebenefits.va.gov/ebenefits/jobs.
4. Military Skills Translator Tools: Use the Military.com Skills Translator to match military skills to civilian job roles: www.military.com/veteran-jobs/skills-translator.
5. Vocational Rehabilitation and Employment (VR&E) Program: Learn about VR&E benefits and services at the VA's VR&E webpage: www.va.gov/careers-employment/vocational-rehabilitation.
6. American Job Centers: Find a local American Job Center near you and access their services at www.careeronestop.org/localhelp/americanjobcenters.
7. Veterans Service Organizations (VSOs):

- The American Legion: www.legion.org
- Disabled American Veterans (DAV): www.dav.org
- Veterans of Foreign Wars (VFW): www.vfw.org

8. Small Business Administration (SBA): Explore resources for veteran entrepreneurs at the SBA's Office of Veterans Business Development website: www.sba.gov/ovbd.
9. GI Bill: Learn about education benefits and access the GI Bill website at www.va.gov/education.
10. Vet Centers: Find local Vet Centers and readjustment counseling services at www.va.gov/find-locations.

11. Give an Hour: Access free mental health services for veterans and their families at www.giveanhour.org.
12. Cohen Veterans Network: Explore mental health clinics and resources at www.cohenveteransnetwork.org.
13. Hire Heroes USA: Access personalized job search assistance, career coaching, and resume support for veterans at www.hireheroesusa.org.
14. Military OneSource: Explore a wide range of resources including education, career, and financial counseling at www.militaryonesource.mil.
15. LinkedIn for Veterans: Leverage the LinkedIn platform's dedicated resources for veterans, including job search tools and networking opportunities at www.linkedin.com/veterans.
16. Troops to Teachers: Discover information and resources for veterans interested in pursuing a career in education at www.proudtoserveagain.com.
17. Veterans Crisis Line: Seek immediate support if in crisis or in need of emotional help by calling the Veterans Crisis Line at 1-800-273-8255 (Press 1) or visit www.veteranscrisisline.net.
18. Military.com Transition Center: Access articles, guides, and resources on transitioning from military to civilian life at www.military.com/transition.
19. Operation REBOOT: Explore career training programs and employment assistance for veterans provided by the US Chamber of Commerce Foundation at www.uschamberfoundation.org/programs/hiring-our-heroes.
20. National Resource Directory: Find comprehensive resources and support across various areas including employment, education, and healthcare at www.nationalresourcedirectory.gov.
21. VA for Vets: Explore career resources, job search tools, and employment support services specifically tailored for veterans at www.vaforvets.va.gov.
22. Military Transition Support Project: Access resources, workshops, and mentoring programs to aid in the transition process at www.mtsp.org.

23. PsychArmor Institute: Gain access to online courses and resources on various topics including employment, education, and mental health at www.psycharmor.org.
24. Military Friendly: Discover a list of military-friendly employers and educational institutions that prioritize hiring and supporting veterans at www.militaryfriendly.com.
25. National Veterans Foundation: Seek assistance, counseling, and resources for veterans through their helpline at 1-888-777-4443 or visit www.nvf.org.
26. VetFran: Explore opportunities for veterans in franchising and access resources to help start a franchise business at www.vetfran.org.
27. Warrior Scholar Project: Access academic boot camps and support for veterans interested in pursuing higher education at www.warrior-scholar.org.
28. Team Rubicon: Engage in disaster response and community service opportunities alongside fellow veterans at www.teamrubiconusa.org.
29. Veterans' Employment and Training Service (VETS): Explore employment resources, training programs, and job search tools provided by the US Department of Labor at www.dol.gov/vets.
30. Operation Homefront: Find support services, financial assistance, and transitional housing for veterans and their families at www.operationhomefront.org.

Printed in the USA
CPSIA information can be obtained
at www.ICGtesting.com
LVHW091804271023
762209LV00052B/910

9 798891 003187